The Azure IoT Handbook

Develop IoT solutions using the intelligent
edge-to-cloud technologies

Dan Clark

<packt>

BIRMINGHAM—MUMBAI

The Azure IoT Handbook

Group Product Manager: Preet Ahuja

Publishing Product Manager: Suwarna Rajput

Book Project Manager: Neil D'mello

Senior Content Development Editor: Adrija Mitra

Technical Editor: Rajat Sharma

Copy Editor: Safis Editing

Proofreader: Safis Editing

Indexer: Pratik Shirodkar

Production Designer: Prashant Ghare

DevRel Marketing Coordinator: Rohan Dobhal

First published: December 2023

Production reference: 1241123

Published by Packt Publishing Ltd.

Grosvenor House

11 St Paul's Square

Birmingham

B3 1RB, UK

ISBN 978-1-83763-361-6

www.packtpub.com

To my wife, Angie, for being so patient with me during these writing adventures.

– Dan Clark

Contributors

About the author

Dan Clark is a senior **business intelligence** (**BI**), data, and programming consultant specializing in Microsoft technologies. Dan is a Microsoft Certified Trainer and past MVP. His current interest is in IoT systems and the potential of the technology. He is focused on learning new technologies and training others on how to best implement them. Dan has published several books and numerous articles on .NET programming, BI development, and now, IoT. He is a regular speaker at various developer and database conferences and user group meetings and enjoys interacting with the Microsoft communities. Previously, Dan was a physics teacher. He is still inspired by the wonder and awe of studying the universe and figuring out why things behave the way they do.

About the reviewers

Balaji M is a Microsoft Azure IoT Specialty-certified IoT engineer who holds a bachelor's in electronics and communications engineering. He has a significant amount of experience in the end-to-end processes of both the hardware and software sections of IoT. His area of expertise is in consumer electronics and automation. Quite recently, he expanded his horizons by stepping into Industry 4.0. His areas of interest include but are not limited to wireless technologies, safety/integrity of systems, low-power devices, data analytics, and ML on edge.

I would like to thank Packt Publishing for providing me with the opportunity to review this excellent book. I would also like to extend my gratitude to my mom and my friends, who have stood beside me always.

Yatish Patil is an author and Azure IoT analytics technology expert with a passion for building customer success consulting with cloud, IoT, and analytics solutions using Microsoft Azure. He has worked with enterprise customers, enabling them to identify and cultivate IoT analytics opportunities through technology innovation.

He is actively involved in technology consulting and solutioning for customers, defining technology roadmaps, best practices, and processes.

He is an ISB product management alumnus who focuses on customer success, solutions, and product management helping customers achieve business goals. He is also the author of *Azure IoT Development Cookbook*, which focuses on the Microsoft Azure IoT platform and preconfigured end-to-end solutions.

I'd like to thank my family and friends, who understand the time and commitment it takes to research and brainstorm constantly changing technology and customer demand. Also, thank you to my teammates and colleagues at various stages, who helped me to gain knowledge with their support.

Table of Contents

Part 1: Capturing Data from Remote Devices

1

An Introduction to the IoT   3

2

Exploring the IoT Hub Service   13

Part 2: Processing the Data

6

7

8

9

Part 3: Processing the Data

10

Visualizing Streaming Data in Power BI   155

11

Integrating Machine Learning   175

12

Responding to Device Events   195

Preface

Welcome to this guide on setting up an **Internet of Things (IoT)** solution in Azure. As the IoT continues to transform the way we interact with the world, IoT developers and architects are at the forefront of innovation. This book is crafted with your specific needs in mind, whether you're a seasoned developer or an architect seeking to navigate the complexities of building and managing IoT solutions in the Azure ecosystem.

With the rise of cloud-based computing, deploying IoT systems has become more cost-effective for businesses. This transformation has led to developers and architects shouldering the responsibility of creating, managing, and securing these systems, even if they are new to the IoT technology. *The Azure IoT Handbook* is a comprehensive introduction to quickly bring you up to speed in this rapidly evolving landscape.

In the world of IoT, building a complete system that operates seamlessly in the Azure cloud is paramount. Our focus is on guiding you through the process of developing a comprehensive IoT system in Azure, from inception to deployment. This book will provide insights into creating, securing, and managing enterprise-wide IoT systems. Additionally, it will equip you with the essential skills to collect, analyze, and visualize streaming data, ensuring that you can harness the full potential of your IoT solutions.

Whether you are embarking on a new IoT project or enhancing your existing skills, this book is here to serve as your trusted companion on your journey toward mastering IoT in Azure. We hope it empowers you to not only navigate the intricacies of IoT development but also to drive innovation, making a meaningful impact in the rapidly evolving IoT landscape.

Starting with the basic building blocks of any IoT system, this book guides you through mobile device management and data collection using an IoT hub. You'll explore essential tools for system security and monitoring. Following data collection, you'll delve into real-time data analytics using Azure Stream Analytics and view real-time streaming on a Power BI dashboard. Packed with real-world examples, this book covers common IoT use cases as well.

By the end of this IoT book, you'll know how to design and develop IoT solutions, leveraging intelligent edge-to-cloud technologies implemented on Azure.

Who this book is for

Our primary audience for this book consists of IoT developers and architects. If you fall into one of these categories and are eager to gain the knowledge and skills required to implement and manage IoT solutions in Azure, you're in the right place. To make the most of this guide, you should possess some prior knowledge of programming languages such as C#, Java, or Python, along with a basic understanding of data processing principles.

What this book covers

Chapter 1, An Introduction to the IoT, introduces you to the exciting world of the IoT. This chapter serves as a foundational overview of IoT, explaining its basic concepts and the significance of connecting everyday objects and devices to the internet. It explores the various applications and potential benefits of IoT, setting the stage for a deeper understanding of this transformative technology.

Chapter 2, Exploring the IoT Hub Service, delves into the core component of IoT systems – the Azure IoT Hub service. This chapter provides a comprehensive exploration of IoT Hub, an essential platform to manage and communicate with IoT devices. You will learn about its key features and functionalities and how it enables secure and scalable IoT solutions. This chapter also introduces important concepts such as telemetry and device management within the context of IoT Hub.

Chapter 3, Provisioning Devices with the Device Provisioning Service, focuses on the crucial aspect of provisioning IoT devices securely. You will gain insights into the Device Provisioning Service, an integral part of IoT architecture. This chapter covers the process of registering and onboarding devices, ensuring that they can seamlessly connect to the IoT ecosystem while maintaining robust security protocols. Understanding device provisioning is essential to build reliable and secure IoT systems, making this chapter a vital resource for IoT enthusiasts and professionals.

Chapter 4, Exploring Device Management and Monitoring, delves into the intricacies of managing and monitoring IoT devices effectively. It covers topics such as device life cycle management, remote configuration, and monitoring device health and performance. You will learn how to ensure the reliability and efficiency of IoT deployments, making this chapter a valuable resource for IoT professionals seeking to maintain and optimize their device ecosystems.

Chapter 5, Securing IoT Systems, addresses one of the most critical aspects of IoT implementation – security. This chapter emphasizes the importance of securing IoT networks and devices against various threats and vulnerabilities. You will explore best practices and strategies to safeguard your IoT systems, including authentication, encryption, and access control. With the increasing concern over IoT security, this chapter provides essential knowledge to build and maintain robust, safe IoT solutions.

Chapter 6, Creating Message Routing, focuses on the efficient and intelligent routing of messages within an IoT ecosystem. This chapter explores the concept of message routing and its significance in ensuring that data is transmitted to the right destinations within an IoT network. You will gain insights into designing and configuring message routing rules, enabling you to effectively manage and process the vast amount of data generated by IoT devices. Understanding message routing is crucial to optimize data flow and enable real-time decision-making in IoT applications.

Chapter 7, Exploring Azure Stream Analytics, dives into the world of real-time data processing and analysis within IoT environments. This chapter explores the capabilities of Azure Stream Analytics, a powerful service provided by Microsoft Azure to ingest, process, and extract valuable insights from streaming data generated by IoT devices. You will learn how to set up and configure Stream Analytics jobs, enabling you to harness the power of real-time data for decision-making and actionable insights in IoT applications.

Chapter 8, Investigating IoT Data with Azure Data Explorer, focuses on the exploration and analysis of historical and large datasets in IoT systems. Azure Data Explorer is a powerful data exploration and visualization service. You will learn how to query and analyze IoT data and gain valuable insights into your data. This chapter equips IoT professionals with the tools to make sense of the wealth of data their devices generate.

Chapter 9, Exploring IoT Edge Computing, delves into the concept of edge computing, an essential component in IoT architecture that enables data processing closer to the source, reducing latency and enhancing efficiency. You will explore the principles of IoT edge computing, understand how to deploy and manage edge devices, and learn how to leverage the advantages of edge computing for real-time processing and decision-making in IoT applications. Understanding edge computing is crucial for creating responsive and scalable IoT systems.

Chapter 10, Visualizing Streaming Data in Power BI, introduces you to the power of real-time data visualization and reporting using Power BI. This chapter explores how to connect and visualize streaming data from IoT devices, providing insights into how to create interactive, dynamic dashboards and reports that display real-time updates.

Chapter 11, Integrating Machine Learning, delves into the convergence of IoT and machine learning. It guides you through the process of leveraging machine learning models to gain deeper insights, predictions, and automation within IoT systems. This chapter covers topics such as model integration, training, and deployment, allowing you to understand how to harness the full potential of machine learning in IoT applications.

Chapter 12, Responding to Device Events, focuses on understanding and managing device events in IoT systems. You will learn how to design responsive and automated workflows that can be triggered by specific device events, allowing real-time actions and decision-making.

To get the most out of this book

To make the most of this guide, you should possess some prior knowledge of programming languages such as C#, Java, or Python, along with a basic understanding of data processing principles. Some experience with the Azure portal and Visual Studio Code would also be helpful.

Software/hardware covered in the book	OS requirements
C# and Python	Windows 10 or above
Visual Studio Code	
The Azure portal	

It is expected that you have access to an Azure subscription. This can be a free version available at `https://azure.microsoft.com/en-us/free/`. It is important to delete your resources after each lab. This will ensure you have enough spend to make it through all the exercises in this book.

If you are using the digital version of this book, we advise you to type the code yourself or access the code from the book's GitHub repository (a link is available in the next section). Doing so will help you avoid any potential errors related to the copying and pasting of code.

Download the example code files

You can download the example code files for this book from GitHub at `https://github.com/PacktPublishing/The-Azure-IoT-Handbook`. This includes chapter code and lab code. If there's an update to the code, it will be updated in the GitHub repository.

> **Important Note**
> Please note that most of the code in the chapters has been written using C#; however, we have also provided the Python version of the source code in the GitHub repository.

We also have other code bundles from our rich catalog of books and videos available at `https://github.com/PacktPublishing/`. Check them out!

Conventions used

There are a number of text conventions used throughout this book.

`Code in text`: Indicates code words in text, database table names, folder names, filenames, file extensions, pathnames, dummy URLs, user input, and Twitter handles. Here is an example: "In Azure, create a resource group called `IoTLab-rg`."

A block of code is set as follows:

```
#r "Newtonsoft.Json"

using System.Net;
using Microsoft.AspNetCore.Mvc;
using Microsoft.Extensions.Primitives;
using Newtonsoft.Json;
```

Any command-line input or output is written as follows:

```
dotnet run --PrimaryConnectionString <myDevicePrimaryConnectionString>
```

Bold: Indicates a new term, an important word, or words that you see on screen. For instance, words in menus or dialog boxes appear in **bold**. Here is an example: "Select the **Overview** page under the **Defender for IoT** group on the IoT Hub left-side menu."

> Tips or important notes
> Appear like this.

Get in touch

Feedback from our readers is always welcome.

General feedback: If you have questions about any aspect of this book, email us at customercare@packtpub.com and mention the book title in the subject of your message.

Errata: Although we have taken every care to ensure the accuracy of our content, mistakes do happen. If you have found a mistake in this book, we would be grateful if you would report this to us. Please visit www.packtpub.com/support/errata and fill in the form.

Piracy: If you come across any illegal copies of our works in any form on the internet, we would be grateful if you would provide us with the location address or website name. Please contact us at copyright@packt.com with a link to the material.

If you are interested in becoming an author: If there is a topic that you have expertise in and you are interested in either writing or contributing to a book, please visit authors.packtpub.com.

Share Your Thoughts

Once you've read *The Azure IoT Handbook,* we'd love to hear your thoughts! Scan the QR code below to go straight to the Amazon review page for this book and share your feedback.

https://packt.link/r/1837633614

Your review is important to us and the tech community and will help us make sure we're delivering excellent quality content.

Download a free PDF copy of this book

Thanks for purchasing this book!

Do you like to read on the go but are unable to carry your print books everywhere? Is your eBook purchase not compatible with the device of your choice?

Don't worry, now with every Packt book you get a DRM-free PDF version of that book at no cost.

Read anywhere, any place, on any device. Search, copy, and paste code from your favorite technical books directly into your application.

The perks don't stop there, you can get exclusive access to discounts, newsletters, and great free content in your inbox daily

Follow these simple steps to get the benefits:

1. Scan the QR code or visit the link below

https://packt.link/free-ebook/978-1-83763-361-6

2. Submit your proof of purchase
3. That's it! We'll send your free PDF and other benefits to your email directly

Part 1: Capturing Data from Remote Devices

Welcome to *The Azure IoT Handbook*, where we explore the world of IoT in a straightforward manner. The first part of the book covers the fundamentals of IoT. We then introduce the core concepts of IoT systems on Azure. This includes the IoT Hub service, followed by the Device Provisioning service, which discusses provisioning devices efficiently and at scale. We will then focus on device management, monitoring, and the importance of securing IoT systems. Whether you're a seasoned IoT professional or a beginner, our goal is to provide you with essential knowledge to navigate the Azure IoT landscape with confidence. Let's embark on this journey together as we delve deeper into the world of IoT.

This part has the following chapters:

- *Chapter 1, An Introduction to the IoT*
- *Chapter 2, Exploring the IoT Hub Service*
- *Chapter 3, Provisioning Devices with the Device Provisioning Service*
- *Chapter 4, Exploring Device Management and Monitoring*
- *Chapter 5, Securing IoT Systems*

1

An Introduction to the IoT

The **Internet of Things** (**IoT**) is rapidly transforming the way we live, work, and interact with the world around us. It has the potential to revolutionize countless industries, from manufacturing and healthcare to transportation and agriculture. With billions of interconnected devices already in use, and that number projected to grow exponentially in the coming years, understanding and harnessing the power of the IoT is becoming increasingly important. This book aims to provide a comprehensive introduction to the IoT and is arranged to guide you through setting up and securing your devices, routing and analyzing data, and then visualizing and augmenting it. Although there are no prerequisites for this book, it would be beneficial to have some experience with programming and scripting languages.

This chapter introduces you to the IoT. It introduces some common use cases for developing IoT systems and introduces you to the common components of an IoT system. You will understand the benefits of an IoT hub and its role in the IoT system. The chapter ends with a hands-on lab where you will connect a virtual IoT device to an Azure IoT hub and capture the streaming data.

In this chapter, we're going to cover the following main topics:

- What is the IoT?
- What are some common IoT use cases?
- What are the components of an IoT system?
- Lab – setting up an IoT hub in Azure

Technical requirement

In order to complete the lab in this chapter, you will need a subscription to Azure. If you do not have access to one, you can sign up for a free trial subscription at `https://azure.microsoft.com/subscription/free`. This free trial will give you $200 of Azure credit to spend over the next 30 days. Take some time to explore all the free services you get with your free subscription; they are quite extensive.

What is the IoT?

If you are reading this book, you are probably somewhat aware of what the IoT is and have been tasked with an IoT project. Nonetheless, it is a good idea to look at the common components of IoT systems. The IoT is a network of connected computing devices with embedded sensors that can transfer the sensor data to the cloud for centralized data processing. These devices run the gamut of wearable devices, such as Fitbit, that track your steps and vitals to smart cars and appliances.

Once you begin planning an IoT solution, you will find that it is a complex task. Both the device side and the cloud side involve complicated implementations that provide hundreds of required features, and the security of these devices is of the utmost importance. This is where Azure comes into play. It provides a set of **Platform-as-a-Service** (**PaaS**) offerings that make creating IoT systems easier, more reliable, and secure.

Now that we know what the IoT is, let's look at some use cases before we dive into further details.

What are some common IoT use cases?

While there are some common IoT use cases to consider, this list is getting bigger all the time. People are constantly coming up with exciting new ways to use the IoT to solve problems, and I think we are just scratching the surface of its full potential. Nevertheless, it is helpful to consider the following common use cases to get a feel for the technology:

- **Agriculture**: This domain is finding many ways to use the IoT. For example, smart irrigation is a popular IoT use case for farmers that provides real-time visibility; IoT sensors recognize when environmental factors, such as ground moisture, call for adjusted watering schedules and can pinpoint where water is needed.

- **Predictive maintenance**: Predictive maintenance sensors and software help recognize when a piece of equipment is out of date, slowing down, or malfunctioning and needs to be replaced proactively instead of reactively.

- **Remote monitoring**: This IoT use case allows companies to monitor sensors that are in dangerous and remote places. This is particularly helpful in industries such as mining and environmental agencies, where specialized equipment is widely distributed.

Now that we have discussed a few use cases, let's review the common components of an IoT system.

What are the components of an IoT system?

At their core, all IoT systems are composed of things, insights, and actions. *Things* refers to small computing devices with embedded sensors that send data up to the cloud through a cloud gateway. The cloud gateway offers a central hub in the cloud that provides secure connectivity, telemetry, event ingestion, cloud-to-device communication, and device management functionalities. It routes the streaming data to storage for long-term analysis and to services that can analyze real-time streaming data. Once the data is processed, it is sent to reporting tools, dashboards, and automated workflows that can automatically respond to the conditions and insights garnered from the telemetry data of the devices. The following figure shows a generic IoT system:

Core Subsystems

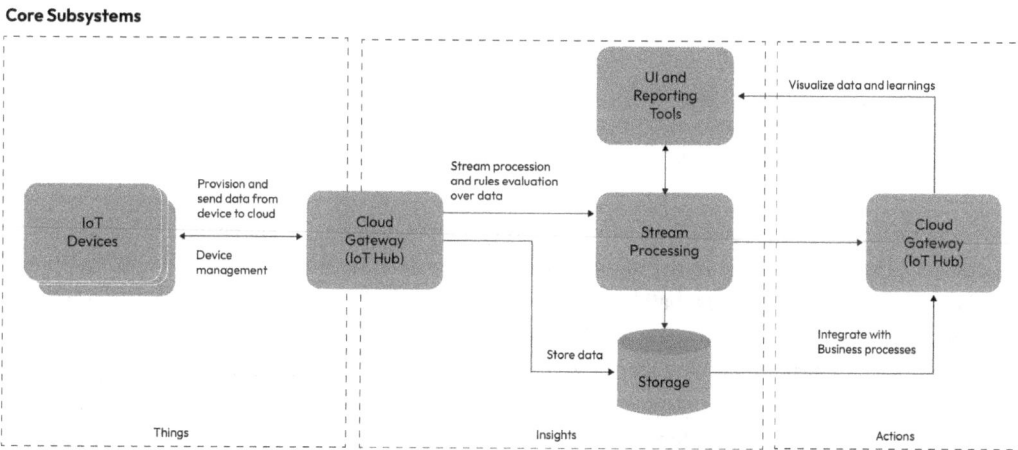

Figure 1.1 – Basic structure of a generic IoT system

In addition to the basic components of an IoT system, many IoT solutions include subsystems for provisioning management at scale, end-to-end security, and machine learning. The following figure shows an IoT system architecture with additional subsystems:

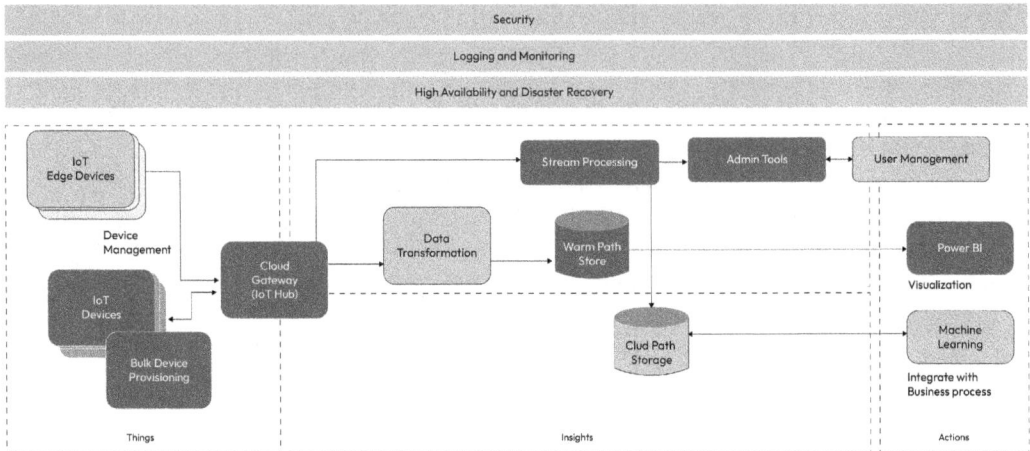

Figure 1.2 – Adding optional subsystems to the IoT solution

As shown in the figure, **Bulk Device Provisioning** allows the provisioning of a large number of devices. Edge devices play an active role in managing access and controlling the flow of information. They can assist in tasks such as filtering, batching, aggregating, and buffering data, as well as translating protocols. **Data Transformation** refers to the manipulation or aggregation of the raw streaming data. The **User Management** subsystem allows the capabilities of different users and groups to be specified in the context of the actions they can perform via the applications and on the systems. **Machine Learning** provides the systems with the ability to perform actions based on the data, such as predictive maintenance and security alerts. In addition to supporting these subsystems, IoT systems should provide the ability to set up system-wide security and end-to-end logging and monitoring, in addition to high availability and disaster recovery for the system.

Now that we have seen the pieces that make up an IoT system, let's see what the corresponding resources are in Azure.

Understanding the Azure IoT system

The first thing we need for an IoT system is the devices that will collect the data. These devices can be very small, such as 8-bit microcontrollers, all the way up to server-grade devices depending on the implementation requirements. There are a tremendous number of devices on the market to choose from. The key differentiators are cost, power, network access, types of sensors, and the inputs and outputs accepted.

Although not necessary, when using an IoT device on Azure it is beneficial to ensure it is an Azure Certified Device. This guarantees the device supports telemetry that will be compatible with the IoT hub and the **Device Provisioning Service** (**DPS**). It will also facilitate cloud-to-device messaging, direct methods, and device-twin updating. Microsoft maintains a catalog of Azure Certified Devices at `https://devicecatalog.azure.com`.

Organizations that choose Azure Certified Devices for their IoT projects benefit from reduced development time, increased security, and a more streamlined integration process when building and scaling their IoT solutions on the Azure platform. These devices are typically used in a wide range of IoT applications, including industrial automation, smart cities, healthcare, agriculture, and more.

After selecting a device, the next step is to set up an IoT hub on Azure and connect the device to it. Within Azure IoT Hub, numerous features are available to simplify device management and control. These features comprise secure communication channels that enhance data transmission and reception, the automatic resending of device messages to accommodate intermittent connectivity, and selective revocation of access rights for specific devices as required.

In the following hands-on lab, you will learn how to set up a virtual device that will send simulated data to an IoT hub you will create in Azure. Make sure you have the technical requirements in place before you start!

Lab – setting up an IoT hub in Azure

By the end of this lab, you will be able to do the following:

- Navigate the Azure portal
- Set up a resource group
- Create an IoT hub
- Create a virtual device
- Send data from the device to the IoT hub

To do this, you will need to perform the following steps:

1. Before you start the lab, you will need to create a unique ID. We will add this to the ends of your resource names to make them unique. This ID can be whatever you want. I find using my initials and then attaching the month and day I was born works well. So I would use `drc0830`.

2. Open a Microsoft Edge browser window, and then navigate to the Azure portal (`https://portal.azure.com`).

3.   Sign in with the credentials you used to create the account. You should land on the home page. You can customize the look and feel by selecting the settings icon in the top-right corner:

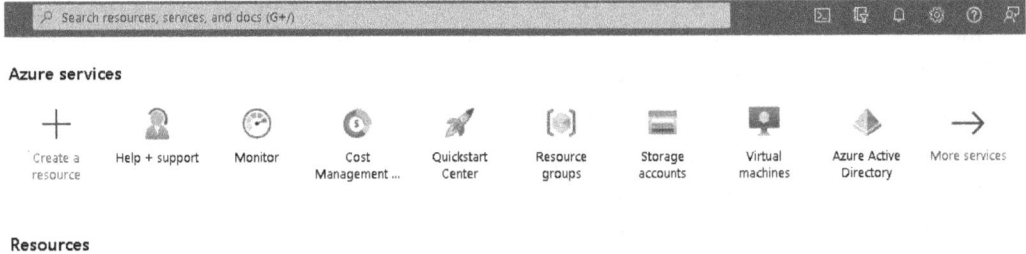

Figure 1.3 – Selecting the settings icon

4.   On the settings page, select the **Appearance + startup views** tab. Make sure the **Menu behavior** option is set to **Docked** (see *Figure 1.4*). Also, verify that the startup page is set to **Home**.

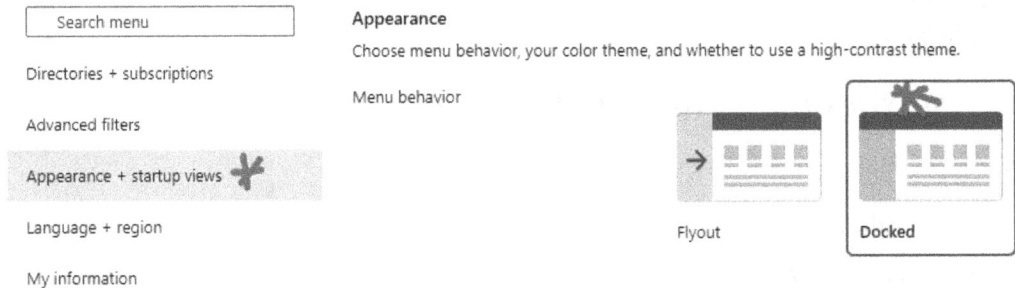

Figure 1.4 – Setting up the look and feel of the portal

5.   Via the left-side menu, go back to the **Home** page.

6.   At the top of the page, you should see a + icon that lets you create a resource. Click on it, and in the search bar that appears, enter `Resource group`, then select **Resource group** from the results dropdown:

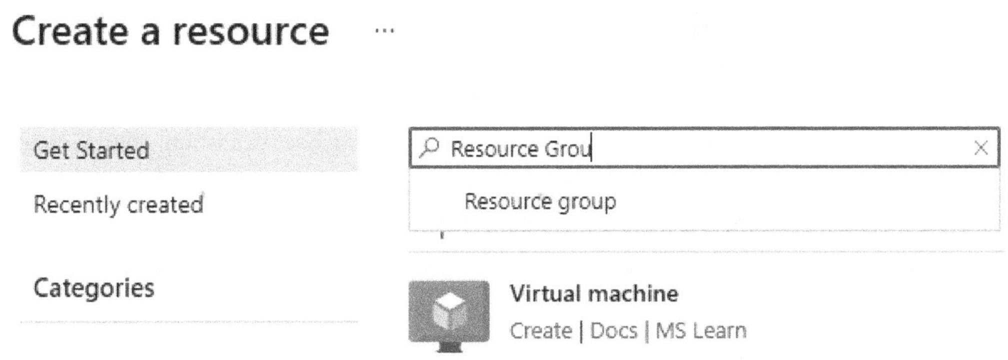

Create a resource   ⋯

Get Started	🔍 Resource Grou ✕
Recently created	Resource group
Categories	▨ **Virtual machine** Create | Docs | MS Learn

Figure 1.5 – Creating a resource group

7. On the next page, click on the **Create** button to start the creation of the resource group.

8. In the **Basics** tab, create a resource group named `rg-iot-training-{your-id}` (replacing `{your-id}` with the one you created earlier). Select a region close to your location and then select **Review + create** at the bottom:

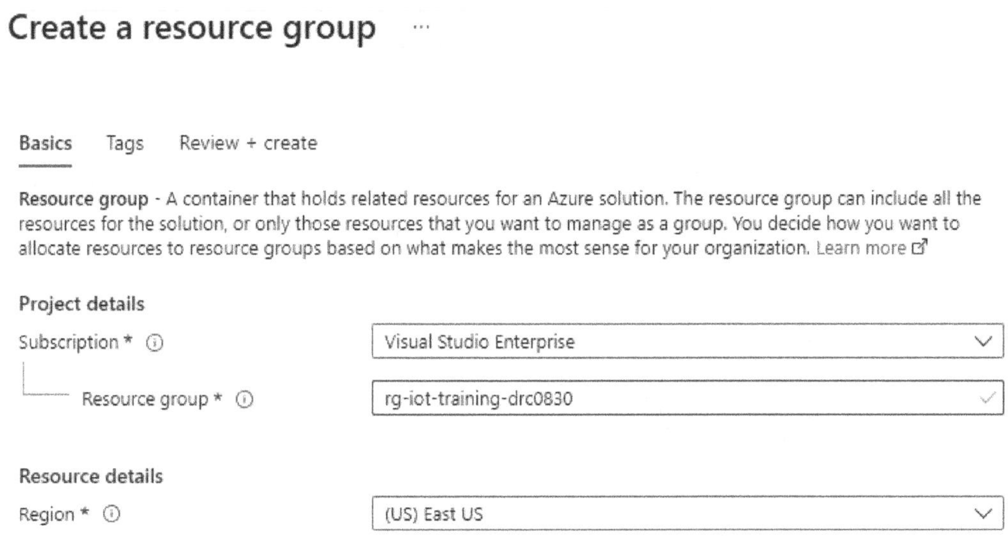

Create a resource group   ⋯

Basics   Tags   Review + create

Resource group - A container that holds related resources for an Azure solution. The resource group can include all the resources for the solution, or only those resources that you want to manage as a group. You decide how you want to allocate resources to resource groups based on what makes the most sense for your organization. Learn more ⬀

Project details

Subscription * ⓘ	Visual Studio Enterprise ⌄
Resource group * ⓘ	rg-iot-training-drc0830 ✓

Resource details

Region * ⓘ	(US) East US ⌄

Figure 1.6 – Creating a resource group

9. Once it passes validation, click on the **Create** button at the bottom of the page.

10. After your new resource group has been created, select **Resource groups** in the menu on the left. You should see the resource group you just created in the list. Select your resource group from this list.

11. At the top of the resource group page, select the **+ Create** button. This brings you back to the Azure Marketplace. Search for and create an IoT hub. Name the hub `iothub-{your-id}`. Select the same region as that of the resource group you created.

12. Select **Review and create** at the bottom of the page. Once it passes validation, select **Create**. Once it is created, click the **Go to resource** button.

13. Investigate the features of the IoT hub using the left-side menu.

14. From the menu on the left, select **Devices**. On the **Devices** page, select **Add Device**.

15. Name the device `device01`, leave the rest of the settings as the defaults, and select **Save**:

Create a device   ···

ℹ️ Find Certified for Azure IoT devices in the Device Catalog

Device ID * ⓘ

`device01`

☐ IoT Edge Device

Authentication type ⓘ

(Symmetric key)   X.509 Self-Signed   X.509 CA Signed

Auto-generate keys ⓘ

☑

Connect this device to an IoT hub ⓘ

(Enable)   Disable

Parent device ⓘ

No parent device
Set a parent device

Figure 1.7 – Creating an IoT device

16. You may have to refresh the page to see the device just created in the device list. Once the device has been created, select it from the list. Copy and save the **Primary connection string** value for the device.

17. In order to simulate a device, we are going to use the Raspberry Pi Azure IoT simulator. You can get access to it by going to `https://azure-samples.github.io/raspberry-pi-web-simulator/`.

18. Go to *line 15* in the code and paste the device connection string you copied earlier:

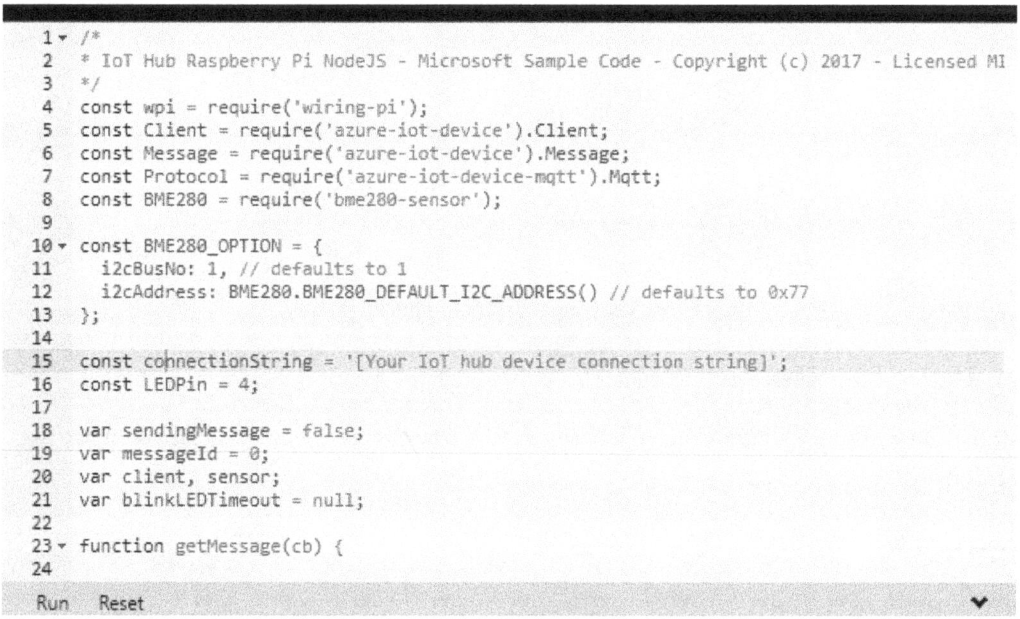

```
1 ▾ /*
2   * IoT Hub Raspberry Pi NodeJS - Microsoft Sample Code - Copyright (c) 2017 - Licensed MI
3   */
4   const wpi = require('wiring-pi');
5   const Client = require('azure-iot-device').Client;
6   const Message = require('azure-iot-device').Message;
7   const Protocol = require('azure-iot-device-mqtt').Mqtt;
8   const BME280 = require('bme280-sensor');
9
10 ▾ const BME280_OPTION = {
11    i2cBusNo: 1, // defaults to 1
12    i2cAddress: BME280.BME280_DEFAULT_I2C_ADDRESS() // defaults to 0x77
13  };
14
15  const connectionString = '[Your IoT hub device connection string]';
16  const LEDPin = 4;
17
18  var sendingMessage = false;
19  var messageId = 0;
20  var client, sensor;
21  var blinkLEDTimeout = null;
22
23 ▾ function getMessage(cb) {
24
Run   Reset                                                                      ▾
```

Figure 1.8 – Updating the connection string

19. Select **Run** and verify messages are being sent.

20. Go back to the Azure portal and navigate to your IoT hub. On the **Overview** page, verify that you are getting signals from the simulated device. From the list labeled **Show data for last**, select **1 Hour**:

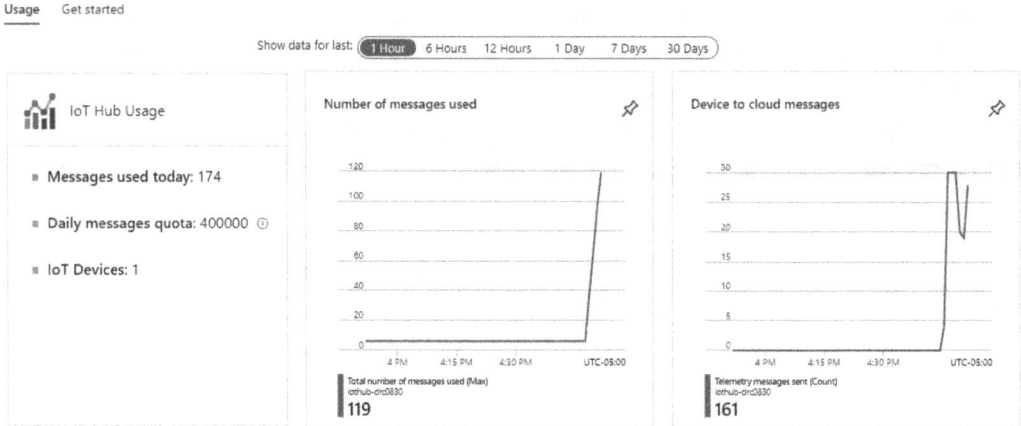

Figure 1.9 – Verifying messages in the IoT hub

21. Once satisfied you are getting signals, stop the simulated device and close the web page.

Now that we are done with this chapter, let's summarize what we have learned so far and where we are going next.

Summary

In this chapter, you learned the basics of an IoT system consisting of things, insights, and actions. You should now understand in a broad sense what kind of services are needed to create an IoT system. In particular, you set up an IoT hub in Azure and verified that it was receiving data from a virtual device. In this chapter, we did not go deep into all the features of an IoT hub. The goal was to get comfortable working in the Azure portal and gain a broad view of the services that make up an IoT system. In *Chapter 2*, *Exploring the IoT Hub Service*, we will take a deeper look at the capabilities of this service.

2

Exploring the IoT Hub Service

In *Chapter 1*, you were given a brief introduction to setting up Azure IoT Hub and using a virtual sensor to send signals to it. Azure IoT Hub is a fully managed service, which makes it easy to set up and secure. It enables dependable and secure two-way communications between IoT devices and a backend solution for collecting and analyzing the data. Most IoT solutions need hundreds and even thousands of sensors. For example, a smart building or smart campus solution needs sensors in every room. When managing these systems at scale, it is very important to have tools that make them more manageable and secure. In this chapter, we will explore some of the features that make Azure IoT Hub an excellent choice for managing and scaling your IoT systems. After completing this chapter, you will have a better understanding of the following areas:

- The features of IoT Hub
- Configuring IoT devices
- Communicating with IoT devices
- Managing and monitoring IoT devices
- Setting up Azure Toolkit in Visual Studio Code
- Lab – working with IoT devices

The features of IoT Hub

You may become confused when you look at the different offerings Microsoft has on Azure. They have **Event Hubs** and **IoT Hub**. *How do you decide when to use which offering?* While they are both message-based service offerings, they are each designed with different use cases in mind. Event Hubs was designed for big data streaming, while IoT Hub was designed to address the unique requirements of addressing and managing IoT devices at scale. IoT Hub uses Event Hubs for its telemetry movement. Some of the advantages of using IoT Hub are as follows:

- IoT Hub offers bi-directional communication between the hub and the devices. This is very important for updating device properties and sending commands to the devices.

- IoT Hub also maintains unique device identities for better security and resilience from attacks.

Now that you have decided to use IoT Hub for your solution, the next step is to *choose the right tier*. IoT Hub has both a **basic tier** and a **standard tier**. To take advantage of cloud-to-device messaging, device twin and device messaging, and IoT edge devices, you need to choose the standard tier. Azure also has a **free tier** for testing and learning experiences that has all the features of the standard tier but is limited by the number of messages it can process and the number of devices you can attach to it.

Each tier is offered in three sizes based on message throughput. The tier 1 limit is up to 1.5 GB/day/unit and 400,00 messages/day/unit, while the tier 2 limit is 22.8 GB and six million, respectively. Tier 3 is a whopping 1,144.4 GB and 300 million, respectively. Since you are charged differently per tier, you must have a good idea of the number and size of the messages being sent.

Once IoT Hub has been set up, the next step is to integrate your IoT devices with it. Let's learn how to do this.

Configuring IoT devices

There are many different **IoT devices** available today. There are devices for consumer IoT, such as home appliances and voice assistance. Commercial IoT devices are used in healthcare and transportation. Industrial IoT is used in manufacturing, while infrastructure IoT is used in smart cities.

What all these devices have in common is that they are hardware devices that collect and exchange data over the internet. They can be standalone devices or embedded devices. When choosing a device, you must consider factors such as power consumption, networking, and available inputs and outputs. You also need to consider security, maintenance, and internet protocols. Fortunately, Microsoft has a certification program that verifies if the device will work on Azure. Bundled solutions are also provided for certain scenarios, such as license plate recognition.

When deploying and communicating with devices it is important that devices have a unique identity and can be authenticated with your message hub in the cloud. Once authenticated, the device needs to send data using a secured protocol. Azure IoT Hub supports several protocols for communication with devices:

- **MQTT**: This is a lightweight publish/subscribe messaging protocol that is commonly used in IoT applications

- **AMQP**: This is a more robust messaging protocol that provides more advanced features, such as guaranteed message delivery and transaction support

- **HTTP**: This is the standard protocol that's used for communication on the World Wide Web, and it can also be used for communication between devices and Azure IoT Hub

- **HTTPS**: This is a secure version of HTTP that uses SSL/TLS to encrypt communication between devices and IoT Hub

- **WebSockets**: This is a protocol that allows for bi-directional communication between a device and IoT Hub over a single connection

Which protocol is best for a particular application will depend on the specific requirements and constraints of the system. Because of its simplicity and low bandwidth requirements, most IoT applications use the MQTT protocol.

You may also need to send commands down to the devices or monitor and set properties on the devices. If you had to create all the code for device communication, security, and maintenance, this would be a complex task. Fortunately, Microsoft has supplied **software development kits** (**SDKs**) that take care of the *plumbing* code for you and allow you to concentrate on the business code. In particular, there is a device SDK that you can install on your device that will facilitate secure communication between Azure IoT Hub and your device. This will be discussed in the following section.

Identity verification

Azure IoT provides various methods of identity verification for IoT devices to ensure secure communication and access control. These methods differ in terms of authentication mechanisms and use cases. Let's take a look at the key identity verification methods:

- **Shared access signatures (SAS tokens)**:

  - **Authentication mechanism**: SAS tokens are generated based on a shared secret (either on the device or service side) and are included in each device's message to authenticate it with IoT Hub.

  - **Use case**: SAS tokens are suitable for devices with limited computing resources as they require minimal processing power for authentication. They are typically used for device-to-cloud communication.

- **X.509 certificates:**

  - **Authentication mechanism**: IoT devices are provisioned with X.509 certificates, typically during manufacturing or deployment. These certificates are used for device authentication when connecting to IoT Hub.

  - **Use case**: X.509 certificates provide a higher level of security and are ideal for devices that require strong authentication. They are often used in scenarios where the device identity needs to be verified without sharing secrets.

- **Device Provisioning Service (DPS):**

  - **Authentication mechanism**: DPS is a service that automates the provisioning of IoT devices, enabling them to join the Azure IoT ecosystem. It uses either SAS tokens or X.509 certificates for authentication during the enrollment process.

  - **Use case**: DPS is particularly useful when dealing with large fleets of IoT devices where manual provisioning is impractical. It ensures that devices can securely and automatically join an IoT solution. (Covered in *Chapter 3*.)

- **Symmetric key authentication:**

  - **Authentication mechanism**: In symmetric key authentication, IoT devices share a secret key with IoT Hub. This shared key is used to authenticate device-to-cloud messages.

  - **Use case**: Symmetric key authentication is suitable for devices that require a simple and efficient method of authentication. However, it is less secure than X.509 certificates.

- **Azure Active Directory (AAD) authentication:**

  - **Authentication mechanism**: AAD authentication allows IoT devices to authenticate using AAD credentials. This method should be used when you want to leverage Azure AD for device identity management.

  - **Use case**: AAD authentication is beneficial when you need to integrate IoT devices with Azure AD for centralized identity management and leverage Azure AD's security features.

The choice of identity verification method depends on factors such as the security requirements of your IoT solution, the capabilities of your IoT devices, and the ease of provisioning and managing devices at scale. It is essential to carefully consider your use case and security needs when selecting the appropriate identity verification method for your Azure IoT devices.

Communicating with IoT devices

There are several different ways you can communicate with your IoT devices. This includes **device-to-cloud messages**, **cloud-to-device messages**, **file upload notifications**, **direct method invocation**, and **operation monitoring events**.

Device-to-cloud messaging

We can use the `Microsoft.Azure.Devices.Client` SDK to create a simulated device. On the device, you will need to install an IoT device SDK and use it to establish a connection to IoT Hub using the connection string you obtained when you registered the device in IoT Hub.

The following code block shows some sample code in Python that demonstrates how to connect a device to IoT Hub using the Azure IoT device SDK:

```python
import os
from azure.iot.device import IoTHubDeviceClient

# Connection string for the device
CONNECTION_STRING = "<your device connection string>"

# Connect to IoT Hub
client = IoTHubDeviceClient.create_from_connection_string(CONNECTION_
STRING)
client.connect()

# Send a message to IoT Hub
message = "Hello, IoT Hub!"
client.send_message(message)
print("Message sent to IoT Hub: {}".format(message))

# Receive messages from IoT Hub
print("Waiting for messages from IoT Hub...")
while True:
    received_message = client.receive_message()
    print("Received message from IoT Hub: {}".format(received_message.
data))
    client.complete_message(received_message)

# Disconnect from IoT Hub
client.disconnect()
```

This code connects to IoT Hub, sends a message, and then disconnects from it.

In addition to sending messages to IoT Hub, you can send direct method calls from the hub to the device.

Using direct methods

In Azure IoT, a direct method is a way for a cloud-based application to remotely invoke a method on a device connected to Azure IoT Hub. This can be used to perform some action on the device or retrieve information from the device. Direct methods are typically used to enable remote management of devices. For example, an administrator might use a direct method to remotely reboot a device or retrieve the current status of the device. Direct methods can also be used to enable devices to receive commands from the cloud, such as to update their firmware or configuration.

To invoke a direct method on a device, the cloud-based application sends a message to the device via Azure IoT Hub, specifying the name of the method to be called and any arguments that need to be passed to the method. The device responds to the message by executing the specified method and returning the result to the cloud-based application. The following code block shows the Python code that's used by the device to create a direct method that sets the time interval of the telemetry being sent:

1. First, it imports the namespaces:

    ```
    import time
    from azure.iot.device import IoTHubDeviceClient, MethodResponse
    ```

2. Next, it defines the direct method:

    ```
    set_telemetry_interval(request, context):
      # Extract the interval value from the request payload
      interval = request.payload

      # Set the telemetry interval
      global TELEMETRY_INTERVAL
      TELEMETRY_INTERVAL = interval

      # Return a response
      return MethodResponse.create_from_method_request(request, 200,
    "Success")
    ```

3. Then, it connects the device client to IoT Hub:

    ```
    client = IoTHubDeviceClient.create_from_connection_
    string(CONNECTION_STRING)
    client.connect()
    ```

4. After, it sets the direct method handler:

```
client.on_method_request_received = set_telemetry_interval

# Wait for direct method calls
while True:
  time.sleep(1)

# Disconnect the device client from the IoT hub
client.disconnect()

# Cloud side
from azure.iot.hub import IoTHubRegistryManager

# Connect to the IoT hub
registry_manager = IoTHubRegistryManager(CONNECTION_STRING)
```

5. Next, it calls the direct method on the device:

```
device_method_params = {"interval": 10}
method_result = registry_manager.invoke_device_method(DEVICE_ID,
"set_telemetry_interval", device_method_params)

# Print the result
print(method_result.status)
print(method_result.payload)
```

The following Python code invokes the direct method call using the SDK to connect to the device and invoke the direct method.

6. Now, it imports the namespaces:

```
import random
import time
import os
from azure.iot.device import IoTHubDeviceClient, Message
```

7. Then, it sets the connection string:

```
CONNECTION_STRING = "<your device connection string>"
```

8. Next, it sets the method name and payload for the direct method:

```
METHOD_NAME = "start"
METHOD_PAYLOAD = {"interval": 10}

def method_request_handler(method_request):
```

```python
    # This function handles incoming direct method requests
    if method_request.name == METHOD_NAME:
        print("Received direct method request: {}".
format(METHOD_NAME))
        response_payload = {"result": "Started"}
        response_status = 200
    else:
        response_payload = {"result": "Unknown method"}
        response_status = 404
```

9. After, it sends a response back to IoT Hub:

```python
    method_response = method_request.create_response(response_
status, response_payload)
    client.send_method_response(method_response)


# Connect to the device
client = IoTHubDeviceClient.create_from_connection_
string(CONNECTION_STRING)


# Set the method request handler
client.on_method_request_received = method_request_handler


# Connect to IoT Hub and wait for incoming direct method
requests
client.connect()
print("Connected to IoT Hub and waiting for direct method
requests...")


while True:
    try:
        time.sleep(1)
    except KeyboardInterrupt:
        print("Disconnecting from IoT Hub...")
        break
```

10. Finally, it disconnects from IoT Hub:

```python
client.disconnect()
```

Another way the device communicates with IoT Hub is through **device twins**. We will discuss these in the next section.

Using a device twin

An **Azure device twin** is a cloud-based representation of a physical device that is connected to Azure IoT Hub. The device twin stores metadata and configuration information for the device in a JSON document format, allowing the device and its cloud-based counterpart to stay in sync. It can be used to query the desired and reported device properties and to set the desired properties for a device. It can be used in several ways:

- **Storing device state information**: The device twin can store the current state of a device, such as its current temperature or whether a valve is open or closed. This information can be used to monitor the device and take appropriate action if necessary.

- **Synchronizing device state**: The device twin can be used to synchronize the desired state of a device with its actual state. For example, if a device's desired state is set to *on*, but its actual state is *off*, the device twin can be used to send a command to the device to turn it on.

- **Sending commands to devices**: The device twin can be used to send commands to a device, such as changing its configuration or updating its firmware.

- **Monitoring device health**: The device twin can be used to monitor the health of a device, such as its battery level or connectivity status. This information can be used to take appropriate action, such as sending a notification if a device's battery is running low or if it has lost connectivity.

A device twin in Azure IoT Hub consists of two main types of properties:

- **Desired properties**: These are the properties that you want a device to have. For example, you might set the desired temperature of a device to be 25 degrees Celsius.

- **Reported properties**: These are the properties that a device reports back to IoT Hub. For example, a device might report its current temperature, battery level, or connectivity status.

The following code block shows an example of a device twin for a device that is used to control the temperature of a room:

```
{
  "deviceId": "temperature-sensor-456",
  "properties": {
    "desired": {
      "samplingInterval": 600
    },
    "reported": {
      "currentTemperature": 70,
      "batteryLevel": 0.95,
      "isConnected": true
    }
  },
```

```
  "tags": {
    "location": "greenhouse"
  },
  "metadata": {
    "type": "temperature sensor"
  }
}
```

In this example, the device is a thermostat with an ID of thermostat-123. The desired properties of the device include the target temperature that the thermostat should try to maintain. The reported properties include the current temperature, battery level, and whether the thermostat is on or off. The device twin also includes metadata about the device, such as its type and location, as well as any tags that have been applied to it.

An IoT device can update its properties from a device twin by using the Azure IoT SDKs. The device can connect to IoT Hub and retrieve its device twin, which includes the desired and reported properties of the device. It can then read the desired properties from the device twin and use them to update its internal state. For example, if the desired temperature for a thermostat is set to 25 degrees Celsius in the device twin, the device can update its internal temperature control settings to try to maintain this temperature.

The device can also report its current state back to the device twin by updating the reported properties in the twin. This can be done by connecting to IoT Hub and sending a message that includes the updated reported properties. IoT Hub will then update the device twin with the new reported properties so that they can be queried by other systems that are connected to the IoT hub.

By updating the desired and reported properties in the device twin in this way, an IoT device can synchronize its state with IoT Hub, and other systems that are connected to IoT Hub can monitor and control the device. The following is some sample C# code for a device that's updating its properties from the device twin using the Azure IoT SDKs:

```csharp
using Microsoft.Azure.Devices;

// Connect to the IoT hub
string connectionString = "HostName=<IoT hub name>.azure-devices.
net;SharedAccessKeyName=<policy name>;SharedAccessKey=<policy key>";
ServiceClient serviceClient = ServiceClient.
CreateFromConnectionString(connectionString);

// Retrieve the device twin for the device
string deviceId = "<device ID>";
Twin twin = await serviceClient.GetTwinAsync(deviceId);

// Read the desired properties from the device twin
double targetTemperature = twin.Properties.
```

```
Desired["targetTemperature"];

// Update the device's internal temperature control settings
UpdateTemperatureControlSettings(targetTemperature);

// Report the current state back to the device twin
var patch = new
{
    currentTemperature = GetCurrentTemperature(),
    batteryLevel = GetBatteryLevel(),
    isOn = IsOn()
};
await serviceClient.UpdateReportedPropertiesAsync(deviceId, patch);
```

In this example, the device first connects to IoT Hub using a connection string that includes the name of the hub and the credentials needed to authenticate with it. It then retrieves its device twin using the `GetTwinAsync` method of the `ServiceClient` class.

Next, the device reads the desired `targetTemperature` property from the device twin and uses it to update its internal temperature control settings. Finally, the device reports its current state back to the device twin by calling the `UpdateReportedPropertiesAsync` method of the `ServiceClient` class and passing it an object that includes the updated reported properties.

Now that we know how to configure and provision IoT devices, let's see how we can manage and monitor the devices. While managing a few devices is not that difficult, it becomes increasingly more difficult as you scale. Fortunately, Azure IoT Hub has a lot of tools you can use to make this a lot easier.

Managing and monitoring IoT devices

To manage IoT devices at scale, we are going to use Azure IoT Hub. Even though we are going to connect only a few devices to Azure IoT Hub, it is very capable of scaling to millions of devices. In this section, you are going to learn how to manage an IoT device using the Azure portal and the Azure **command-line interface (CLI)**.

When managing devices, there are five stages that you need to be aware of:

- **Planning** consists of creating a metadata schema that allows you to query devices and complete bulk management operations. Azure IoT Hub uses device twins to store the metadata in JSON format.

- **Provisioning** is the process of securely provisioning new devices either individually or in bulk and immediately reporting their condition and capabilities.

- **Configuring** is the process of updating device properties and firmware updates.

- **Monitoring** involves observing the overall health and sending alerts to operators when attention is needed. This also includes monitoring the devices for any security threats that may occur.

- **Retirement** involves decommissioning the device at the end of service.

One important aspect of managing IoT devices is **device connectivity**. It is important to know whether a device is connected or not. We can set up **Azure Monitor** to create alerts and notifications based on metric thresholds, which can be used to alert you when a device becomes disconnected. You can also use Azure Monitor to create a log query to retrieve the connection status of all devices in your IoT Hub. To do this, you will need to perform the following steps:

1. First, you need to enable diagnostic logging for your IoT Hub.

2. Once this is enabled, you can go to the **Monitoring** section from the left-hand side menu for your hub and select **Logs**. This will allow you to query the logs using KQL. The following figure shows the query that we used to select all devices that connected in the last 24 hours:

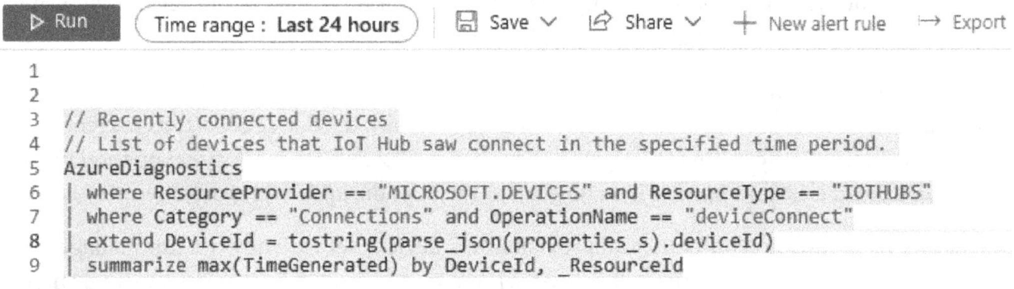

```
▷ Run      ( Time range : Last 24 hours )      🖫 Save ∨    ◪ Share ∨    + New alert rule    ↦ Export

1
2
3  // Recently connected devices
4  // List of devices that IoT Hub saw connect in the specified time period.
5  AzureDiagnostics
6  | where ResourceProvider == "MICROSOFT.DEVICES" and ResourceType == "IOTHUBS"
7  | where Category == "Connections" and OperationName == "deviceConnect"
8  | extend DeviceId = tostring(parse_json(properties_s).deviceId)
9  | summarize max(TimeGenerated) by DeviceId, _ResourceId
```

Figure 2.1 – Querying the log files

3. You can also look at a common set of metrics. For example, using metrics, we can see a chart of the number of devices that are connected:

+ New chart   ◯ Refresh   ⤳ Share ∨   ☺ Feedback ∨

Local Time: Last 30 minutes (Automatic - 1 minu...

Avg Connected devices for iothubdrc1963 ✎

⤳ Add metric   ˚▼ Add filter   ≈ Line chart ∨   ⤵ Drill into Logs ∨   ⯐ New alert rule

✕ Apply splitting   💾 Save to dashboard∨   ...

⤳ iothubdrc1963, **Connected devices**, Avg ⊗

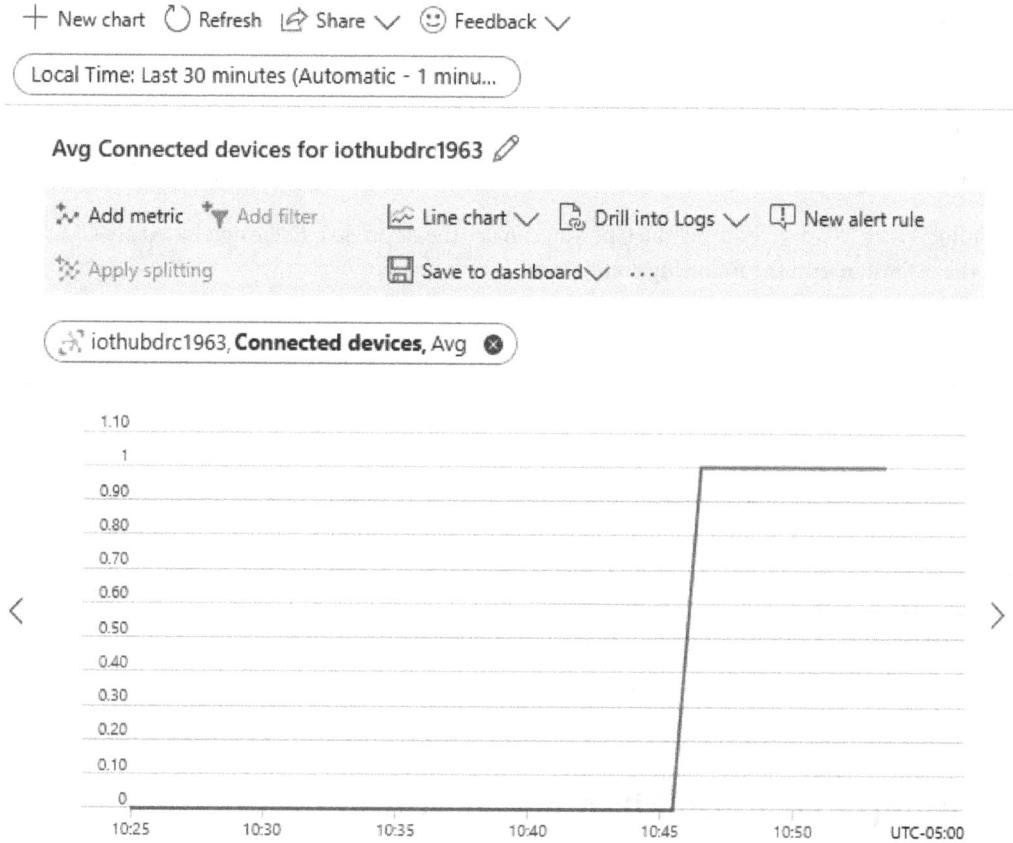

Figure 2.2 – Viewing the results of the query

In addition to the Azure portal, you can use the Azure CLI to manage and monitor your devices. You will learn how to do this in the next section.

Using the Azure CLI to monitor devices

The Azure CLI is a set of command-line tools that allow you to manage Azure resources. It provides an easy way to create, configure, and manage Azure resources from the command line or in scripts. The Azure CLI is available on Windows, Linux, and macOS, and you can use it with **Azure Cloud Shell**, a web-based shell that you can use to manage Azure resources from a web browser. Azure Cloud Shell is also integrated within the Azure portal.

With the Azure CLI, you can perform a wide range of tasks, including the following:

- Creating and managing Azure resources such as virtual machines, storage accounts, and web apps
- Monitoring resource usage and diagnosing issues
- Automating deployments and resource management tasks

To monitor Azure IoT devices using the CLI, you can use the Azure IoT Extension for Azure CLI. To install the extension, run the following command:

```
az extension add --name azure-iot
```

Once the extension has been installed, you can use it to manage your IoT devices and resources. For example, you can use the following command to list all the IoT devices registered in your hub:

```
az iot hub device-identity list --hub-name {YourIoTHubName}
```

You can also use the CLI to send messages to your IoT devices, monitor device-to-cloud and cloud-to-device messages, and more. For more information on using the Azure IoT Extension for Azure CLI, you can refer to the following documentation: `https://docs.microsoft.com/en-us/cli/azure/ext/azure-iot/iot?view=azure-cli-latest`.

To conclude this chapter on IoT devices, you will complete a hands-on lab to gain some experience working with an IoT device. But before that, let's learn how to set up Azure Toolkit.

Setting up Azure Toolkit in Visual Studio Code

Visual Studio Code (**VS Code**) is the preferred **integrated development environment** (**IDE**) and there is an extension called Azure Toolkit that makes it easy to work with and create services in Azure. Although we will be doing most of our work through the portal, you may want to install and investigate using Azure Toolkit in VS Code.

Setting up Azure Toolkit in VS Code allows you to work with Azure services and resources directly from your code editor. Here's a step-by-step guide to set up Azure Toolkit in VS Code:

1. **Install VS Code**: If you haven't already, download and install VS Code from the official website: `https://code.visualstudio.com/`.
2. **Install the Azure Account extension**:
    I.   Open Visual Studio Code.
    II.   Go to the **Extensions** view by clicking on the **Extensions** icon in the sidebar or by pressing *Ctrl + Shift + X* (*Cmd + Shift + X* on macOS).
    III. Search for `Azure Account` in the **Extensions** search bar.
    IV.   Click the **Install** button to install the Azure Account extension.

3. **Sign in to your Azure account**:

    I.   After installing the Azure Account extension, you'll see an Azure icon in the activity bar on the side.

    II.   Click on the Azure icon to open the **Azure** pane.

    III.   In the **Azure** pane, you'll see options to sign in. Click on **Sign in to Azure...** and follow the prompts to sign in to your Azure account.

4. **Install Azure Toolkit for VS Code**:

    I.   Once you've signed in, you can search for `Azure Toolkit for Visual Studio Code`

    II.   Click the **Install** button to install the Azure Toolkit extension.

5. **Access Azure resources**:

    I.   After installing Azure Toolkit, you'll have access to various Azure services and resources directly within VS Code.

    II.   You can use the Azure activity bar to browse and interact with your Azure resources, create and manage Azure resources, deploy applications, and more.

6. **Create and deploy resources**:

    I.   You can use **Azure Resource Manager (ARM)** templates to define and deploy Azure resources.

    II.   You can create, modify, and deploy these templates directly from VS Code using Azure Toolkit.

7. **Working with Azure Functions, Web Apps, and other Azure services**: Azure Toolkit provides extensions for specific Azure services. For example, if you're working with Azure Functions or Azure Web Apps, you can install additional extensions for these services.

8. **Configure your environment**: It's essential to configure your Azure environment properly, including selecting the Azure subscription you want to work with.

9. **Access the Azure CLI**: Azure Toolkit integrates with the Azure CLI, which allows you to run Azure CLI commands within VS Code.

10. **Develop and debug Azure apps**: You can use VS Code's built-in debugging capabilities and extensions to develop and debug Azure applications.

Remember to refer to the official Azure Toolkit documentation for more details and updates as the setup process and features may evolve.

With that, we can start our lab!

Lab – setting up and monitoring IoT devices

After completing this lab, you will be able to do the following:

- Set up a virtual device using VS Code and the IoT SDKs

- Create a direct method on a device and call it from IoT Hub

- Monitor a device using the CLI

To do this, you will need to perform the following steps:

1. Open VS Code and using *Control + Shift + P*, open the Command Palette.

2. Enter the `Azure IoT Hub: Create Device` command in the Command Palette.

3. Enter the device ID as `Device02`.

4. Once the device has been created, you should see the connection string for the device. Copy this for later.

5. Press the *Ctrl + ~* keys to open the terminal.

6. Run the following code in the terminal to create a new project and install the SDK:

```
mkdir Device02
Cd Device02
dotnet new console
dotnet add package Microsoft.Azure.Devices.Client
```

7. Open the `Device02` folder in VS Code and open the `Program.cs` file.

8. Replace the code inside that file with the following:

```
using System;
using System.Text;
using System.Threading.Tasks;
using Microsoft.Azure.Devices.Client;
using Newtonsoft.Json;

namespace Device02
{
    class Program
    {
        private static DeviceClient? deviceClient;
        private readonly static string connectionString = "";
        private static void Main(string[] args)
        {
            Console.WriteLine("IoT Hub C# Simulated Device.
Ctrl-C to exit.\n");
```

```
            deviceClient = DeviceClient.
CreateFromConnectionString(connectionString, TransportType.
Mqtt);
            SendDeviceToCloudMessagesAsync();
            Console.ReadLine();
        }
        private static async void
SendDeviceToCloudMessagesAsync()
        {
            // Create an instance of our sensor
            var sensor = new EnvironmentSensor();
            while (true)
            {
                // read data from the sensor
                var currentTemperature = sensor.
ReadTemperature();
                var currentHumidity = sensor.ReadHumidity();
                var messageString =
CreateMessageString(currentTemperature, currentHumidity);
                // create a byte array from the message string
using ASCII encoding
                var message = new Message(Encoding.ASCII.
GetBytes(messageString));
                // Add a custom application property to the
message.
                // An IoT hub can filter on these properties
without access to the message body.
                message.Properties.Add("temperatureAlert",
(currentTemperature > 30) ? "true" : "false");
                // Send the telemetry message
                await deviceClient.SendEventAsync(message);
                Console.WriteLine("{0} > Sending message: {1}",
DateTime.Now, messageString);
                await Task.Delay(1000);
            }
        }
        private static string CreateMessageString(double
temperature, double humidity)
        {
            // Create an anonymous object that matches the data
structure we wish to send
            var telemetryDataPoint = new
            {
                temperature = temperature,
                humidity = humidity
            };
```

```csharp
            // Create a JSON string from the anonymous object
            return JsonConvert.
SerializeObject(telemetryDataPoint);
        }
    }

    internal class EnvironmentSensor
    {
        // Initial telemetry values
        double minTemperature = 20;
        double minHumidity = 60;
        Random rand = new Random();
        internal EnvironmentSensor()
        {
            // device initialization could occur here
        }
        internal double ReadTemperature()
        {
            return minTemperature + rand.NextDouble() * 15;
        }
        internal double ReadHumidity()
        {
            return minHumidity + rand.NextDouble() * 20;
        }
    }
}
```

This code simulates a device that sends temperature and humidity values to IoT Hub.

9. On *line 12*, enter the connection string you copied in *step 4*.

10. Right-click the `Program.cs` file in the **Explore** window and select **Open** in **Integrated Terminal**. In **Integrated Terminal**, run `Dotnet run`.

11. You should see sample temperature and humidity readings being written to the terminal window. Leave the simulator running and open the Azure portal in your web browser.

12. Open your hub in the portal, select the **Overview** page, and verify that messages are being sent.

13. You can use the Azure CLI to monitor the device streams coming into your hub. Open Cloud Shell in the portal and make sure you select **Bash** from the left-hand dropdown, not **Powershell**. Enter the following commands, making sure you use your IoT Hub name:

```
az extension add --name azure-iot
az iot hub monitor-events  --hub-name myIoTHub
```

You should see the messages being received by the hub. Press *Ctrl + C* to stop the monitoring process.

14. Next, you need to add a direct method to the device. In VS Code, press *Ctrl + C* in the terminal to stop the virtual device.

15. Replace the code in the `Main` method with the following code to set up the callback method handler for the direct method call to the main method. Add this to the code around *line 15*:

```
Console.WriteLine("IoT Hub C# Simulated Device. Ctrl-C to
exit.\n");
deviceClient = DeviceClient.
CreateFromConnectionString(connectionString, TransportType.
Mqtt);
deviceClient.SetMethodHandlerAsync("reboot", onReboot, null).
Wait();
SendDeviceToCloudMessagesAsync();
Console.ReadLine();
```

16. Add the following code to implement a simulated reboot method in the device. This should be added around *line 21*:

```
private static Task<MethodResponse> onReboot(MethodRequest
methodRequest, object userContext)
{
    try
    {
        Console.WriteLine("Rebooting!");
    }
    catch (Exception ex)
    {
        Console.WriteLine();
        Console.WriteLine("Error in sample:    {0}",ex.Message);
    }
    string result = @"{""result"":""Reboot started.""}";
return Task.FromResult(new MethodResponse(Encoding.UTF8.
GetBytes(result), 200));
}
```

17. To test the direct method, you can call it from the portal. Go to the **IoT Hub** page and select the **Devices** tab under the **Device Management** section of the left-hand menu. Select `Device02`. At the top of the **Device02** page, you should see a **Direct method** tab. On that tab, provide the method name, as shown in the following figure, and invoke the method. You should see the result with a status of **200** and a payload of **Reboot started**:

Direct method   ...
device02

Device ID

> device02   📋

Method name * ⓘ

> reboot

Payload ⓘ

>

Response timeout ⓘ   Connection timeout ⓘ

> 30 seconds   ∨   Device must already be connected ∨

Invoke method

Result

```
{
    "status": 200,
    "payload": {
        "result": "Reboot started."
    }
}
```

Figure 2.3 – Invoking a direct method from the Azure portal

Now that you have gained hands-on experience working with devices and communicating between IoT Hub and its devices, let's see where we are going next.

Summary

In this chapter, we delved into the exciting world of IoT and explored various key aspects related to IoT Hub management and device communication.

We began by examining the essential features of IoT Hub, which serves as a central component in IoT solutions. You learned about its role in managing and facilitating communication between IoT devices, making it a crucial part of any IoT ecosystem.

Next, we delved into the process of configuring IoT devices. You gained insights into how to set up and prepare IoT devices for integration with IoT Hub, ensuring they are ready to send and receive data seamlessly.

Communication lies at the heart of IoT, and we explored the various methods and protocols for effectively communicating with IoT devices. You learned how to establish bidirectional communication channels to exchange data with these devices securely.

IoT device management is a critical aspect of maintaining an IoT ecosystem. We discussed strategies and tools for managing and monitoring IoT devices efficiently, including techniques for remote updates and troubleshooting.

To reinforce your understanding, we provided a hands-on lab experience. In this practical exercise, you had the opportunity to apply the knowledge you gained throughout this chapter by working with simulated IoT devices, configuring them, establishing communication, and managing them within an IoT Hub environment.

In the next chapter, you will look at **Device Provisioning Service (DPS)**. In particular, we will discuss Azure DPS, a cloud-based service provided by Microsoft Azure that enables the automatic registration and provisioning of devices to an IoT solution.

3

Provisioning Devices with the Device Provisioning Service

Azure **Device Provisioning Service (DPS)** is a cloud-based service provided by Microsoft Azure that enables the automatic registration and provisioning of devices to an IoT solution. Azure DPS simplifies the task of configuring and deploying IoT devices at scale. With Azure DPS, developers can easily manage the entire life cycle of their IoT devices, from initial provisioning to updates and maintenance.

The service works by providing a secure and scalable way to authenticate and authorize devices before they can connect to an IoT hub. Azure DPS uses industry-standard security protocols such as X.509 certificates and symmetric keys to provide secure communication between devices and the cloud.

In this chapter, you will discover the features of DPS and how to use it to automate your device provisioning. We will be covering the following key topics:

- Device provisioning at scale
- Managing device provisioning security concerns
- Allocation policies in DPS
- Deprovisioning and disenrolling devices
- Using device provisioning SDKs
- Lab – provisioning devices using DPS
- Lab – using custom policies for device allocation

Device provisioning at scale

As you saw in *Chapter 2*, device provisioning is fairly easy when you are provisioning a few devices through IoT Hub. You just need to provide a unique ID and some kind of **attestation**. Attestation is proof that you are who you say you are and can be implemented using a key or certificate. Once a connection has been established, it is configured to its initial state. If this process is manually repeated for many devices, it becomes error-prone and inefficient.

Fortunately, Microsoft has provided DPS to make this process more efficient and reliable. It includes features such as secure attestation, enrollment lists, allocation policies, monitoring, logging, encryption, multi-hub support, and cross-regional support. It also supports open source SDKs in various languages to program against its APIs.

To get started with DPS, you can provision one through the Azure portal or use the CLI. The following figure shows how to provision DPS in the Azure portal:

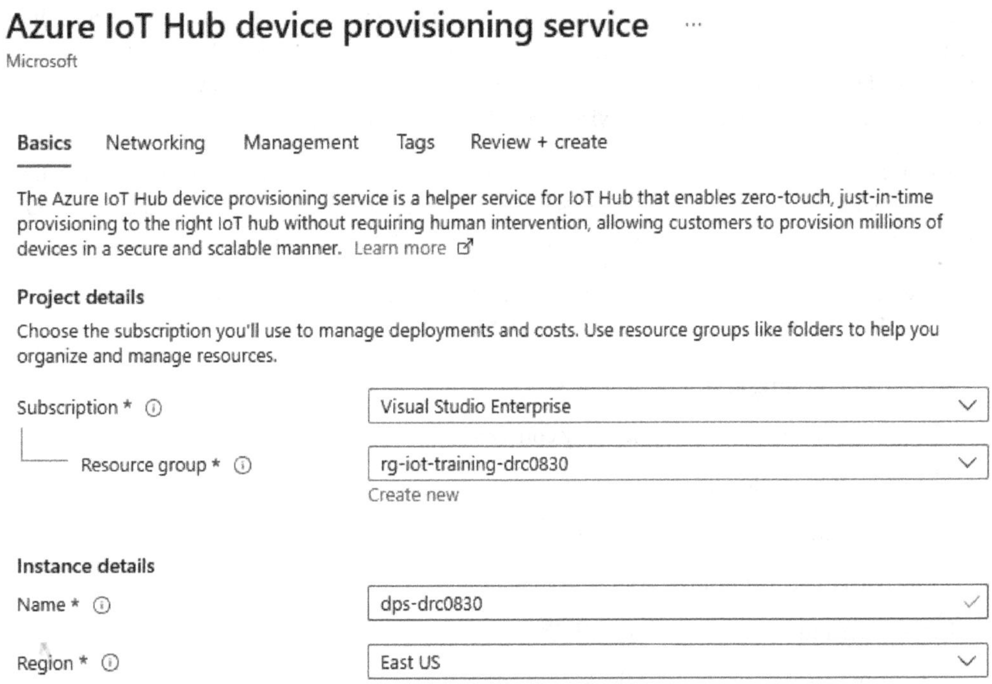

Azure IoT Hub device provisioning service ···
Microsoft

Basics   Networking   Management   Tags   Review + create

The Azure IoT Hub device provisioning service is a helper service for IoT Hub that enables zero-touch, just-in-time provisioning to the right IoT hub without requiring human intervention, allowing customers to provision millions of devices in a secure and scalable manner. Learn more

Project details

Choose the subscription you'll use to manage deployments and costs. Use resource groups like folders to help you organize and manage resources.

Subscription * ⓘ   Visual Studio Enterprise

└── Resource group * ⓘ   rg-iot-training-drc0830
Create new

Instance details

Name * ⓘ   dps-drc0830

Region * ⓘ   East US

Figure 3.1 – Provisioning a DPS service

Once DPS has been provisioned in the **Settings** tab on the left-hand side of the portal, you can link it to one or more IoT hubs. DPS is a helper service that automates the deployment process to reduce the time and risks involved in deploying massive amounts of IoT devices. The following figure shows the steps needed to provision the device:

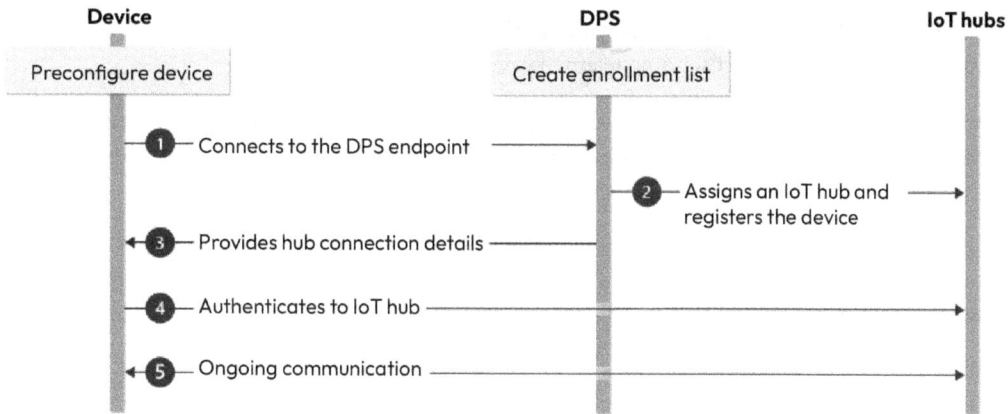

Figure 3.2 – Steps to provision an IoT device with DPS

Before provisioning the device, the device manufacturer must preconfigure the device with its authentication credentials, a DPS ID, and an endpoint. DPS must be pre-configured with an enrollment list that defines valid devices and how they should be provisioned. We will discuss this in a lab at the end of this chapter.

Once the device and cloud have been set up for provisioning, it goes through a provisioning process that is designed to automate the registration process with Azure IoT Hub. The following steps are involved in this process:

1. **The device connects to a network**: The first step is for the device to connect to a network, either through Wi-Fi, Ethernet, or another method.

2. **The device initializes and starts running**: Once the device is connected to the network, it initializes and starts running its firmware or operating system.

3. **The device queries DPS**: During the initialization process, the device queries the Azure DPS instance to obtain the information required for registration, such as the registration ID scope and authentication credentials.

4. **The device authenticates with DPS**: The device uses the authentication credentials provided by DPS to authenticate itself. This ensures that the device is authorized to connect to DPS and register with Azure IoT Hub.

5. **DPS authorizes the device**: DPS verifies the authentication credentials provided by the device and authorizes it to register with Azure IoT Hub.

6. **The device registers with Azure IoT Hub**: Once the device has been authorized, it registers with Azure IoT Hub and receives the necessary configuration information, such as the device twin and device configuration.

7.   **The device starts sending telemetry data**: With the necessary configuration information in hand, the device can start sending telemetry data to Azure IoT Hub.

The types of enrollment supported by DPS are **individual enrollment** and **group enrollment**. Individual enrollments are for devices that have unique initial configurations and can use either X.509 leaf certificates or SAS tokens for attestation.

Group enrollments are for a group of devices that share a specific attestation mechanism (either a certificate or symmetric key) and have the same initial configuration. The following screenshot shows the process of adding an individual enrollment for a device:

Figure 3.3 – Setting up an individual enrollment

In this case, it is using an X.509 certificate, but you can also use **Trusted Platform Module** (**TPM**) or a **symmetric key**. It also allows you to select how you want to allocate devices to your IoT hubs. You can choose between the lowest latency, evenly distributed, a static configuration, or custom using an Azure function. At the bottom of the page, you can create the JSON definition for **Initial Device Twin State**. Creating the group enrollment is very similar, except you do not have the option of the TPM attestation.

Now, let's look at the security concerns when provisioning devices.

Managing device provisioning security concerns

There are many IoT device provisioning security concerns that must be managed. However, DPS offers a comprehensive solution to manage and secure device provisioning in IoT scenarios:

- **Secure boot**:

  - **Concern**: Ensuring that only authenticated and authorized firmware can be run on the device during bootup

  - **Solution**: DPS supports secure boot using X.509 certificates, which ensures that only authenticated and authorized firmware can be run on the device during bootup

- **Strong device authentication**:

  - **Concern**: Using secure methods such as certificates or secure key pairs to authenticate devices before allowing them to connect to a network or system

  - **Solution**: Azure DPS uses X.509 certificates for device authentication and can also integrate with Azure **Active Directory** (**AD**) for authentication and authorization

- **Device management**:

  - **Concern**: Implementing a system for remotely managing and updating devices, including the ability to revoke access or push security updates

  - **Solution**: DPS includes a device management feature that allows you to remotely manage and update devices, including the ability to revoke access or push security updates

- **Network segmentation**:

  - **Concern**: Segregating IoT devices on their own network segment, to limit their access and potential impact on the rest of the network

  - **Solution**: DPS allows you to segment devices into different groups, which enables you to limit their access and potential impact on the rest of the network

- **Regular vulnerability scanning**:

  - **Concern**: Regularly scanning for vulnerabilities on all connected devices and applying any security updates as soon as they are available

  - **Solution**: Azure Security Center can be integrated with DPS for regular vulnerability scanning and remediation of connected devices

- **Encryption**:

  - **Concern**: Using encryption to secure data in transit and at rest

  - **Solution**: DPS supports encryption for data in transit and at rest

- **Regular monitoring**:

  - **Concern**: Regularly monitoring devices and networks for unusual activity and taking appropriate action when necessary

  - **Solution**: DPS includes built-in monitoring and logging features that allow you to track device provisioning events and detect any unusual activity

- **Access control**:

  - **Concern**: Identity and access management

  - **Solution**: DPS integrates with Azure AD for access control and allows you to define roles and permissions for different users and devices

Along with managing the security concerns when provisioning at scale, DPS allows you to allocate devices to various IoT hubs, depending on policies. Let's take a look at how this is done.

Allocation policies in DPS

Allocation policies in DPS determine how devices are assigned to IoT Hub. There are four policies you can choose from:

- **Evenly weighted distribution**: The policy uses a weighted hash where all IoT hubs have the same weight assigned to them. You can adjust the weighted hash of a hub so that it gets more or fewer devices allocated to it.

- **Lowest latency**: Devices are assigned according to the shortest communication time with the IoT device.

- **Static configuration**: Devices are provisioned to a hub specified in the enrolment.

- **Custom, using an Azure function**: This option provides any type of policy in a function that uses a custom webhook that gets called when provisioning the device.

The default policy is **Evenly weighted distribution**. The following figure shows how to set the allocation policy in the portal:

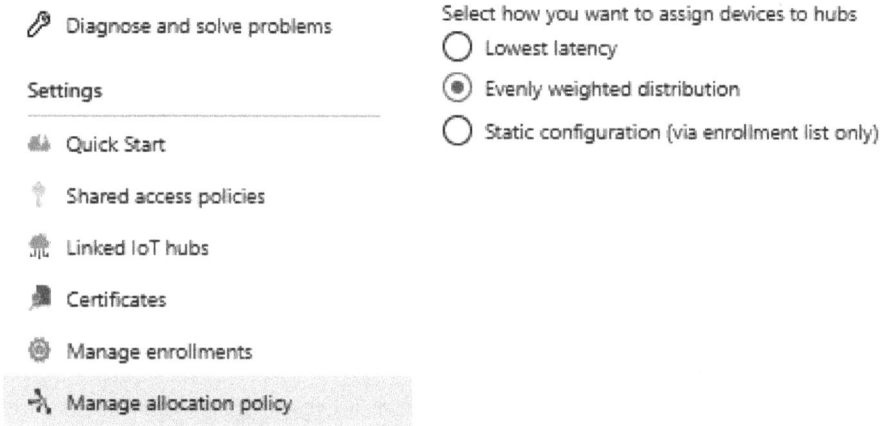

Figure 3.4 – Setting the allocation policy

A common practice when working with IoT devices is the need to reprovision devices. Some of the reasons for reprovisioning an IoT device to a different hub could be to improve latency via geolocation or repurposing the device to a new solution. When a device is reprovisioned, you need to decide what to do with the device's state (composed of the device twin and the device's capabilities). Over time, the device state on the hub will most likely change, while the device state held by DPS is the initial state of the device when it was provisioned. This allows us to either reprovision a device and migrate the current device state or reset it to its initial state.

Deprovisioning and disenrolling devices

When a device has reached the end of its lifetime, we need to deprovision it. Deprovisioning is a two-step process:

1. First, it needs to be disenrolled from DPS; then, you need to deregister the device in IoT Hub.

2. Next, the device needs to be deleted or disabled:

    *   For **individual enrollments**, delete or disable the device on the device provisioning service. If the device uses an X.509 certificate and an enrollment group exists for a signing certificate in that device's certificate chain, the device can be re-enrolled. Next, delete or disable the device in the identity in IoT Hub.

- If the device was provisioned through an **enrollment group**, disable the device in DPS. Disabling the device revokes access but allows other devices in the group access. Do not delete the device. Deleting the device from a group will allow the device to re-enroll with the group. Once the device has been disabled, you can use the list of provisioned devices for the group to find the hub it was provisioned to, as shown in the following screenshot:

Home > dps-drc0830 | Manage enrollments >

group1   ⋯
Enrollment Group Details

↻ Refresh   ✕ Delete Registrations

Settings   **Registration Records**

ⓘ You can view devices that have provisioned via this enrollment group and remove the registration records for previously

🔍 Search devices in this enrollment group

Device Id	↑↓	Assigned IoT Hub
☐ device02		drciothub.azure-devices.net
☐ device01		iothub-drc0830.azure-devices.net

Figure 3.5 – Getting the enrollment list

Although registering a few devices through the portal is a good way to learn the process for production, you will most likely use code and the SDK because this is often more convenient. In the following section, we will look at how to use the SDK.

Using the device provisioning SDK

Although working in the Azure portal is a good way to learn how to set up a DPS service and register devices, when working with a large number of devices and performing automation, it is more convenient to use code and the various SDKs that are available. There is an SDK you can use for each programming language (C, C#, Python, and Java). To provision a device using the SDK, first, create your IoT Hub and your DPS and link your IoT Hub to your DPS. You can do this using the Azure portal, the CLI, or any of the supported language's SDKs. The following code snippet uses the `Microsoft.Azure.Devices` SDK to provision a device using symmetric key attestation:

```
var security = new SecurityProviderSymmetricKey(registrationId,
                              individualEnrollmentPrimaryKey,
                              individualEnrollmentSecondaryKey)
```

```
var transport = new
ProvisioningTransportHandlerAmqp(TransportFallbackType.TcpOnly)
deviceClient = await ProvisionDevice(provClient, security)
```

You will get a chance to use the SDK in the following hands-on lab.

Lab – provisioning devices using DPS

Let's say you have a fleet of refrigerated trucks you need to track. For example, you need to know the location, temperature, and humidity of the trucks. In the context of this use case, you will learn how to do the following:

- Create a new individual enrollment in DPS
- Create a group enrollment

Let's get started.

Adding an individual enrollment

The following steps describe how we can create an individual enrollment in DPS:

1. In Azure, create a resource group. In the resource group, create an IoT Hub and a DPS service.

2. In the DPS service, link to IoT Hub.

3. Open your DPS service in the Azure portal. In the left-hand side menu, locate **Manage enrollments** under the **Settings** header.

4. On the **Manage enrollments** page, select **Add individual enrollments**. Use **Symmetric Key** under **Mechanism**. Select **Auto-generate keys** and specify device01 for **Registration ID** and **IoT Hub Device ID**:

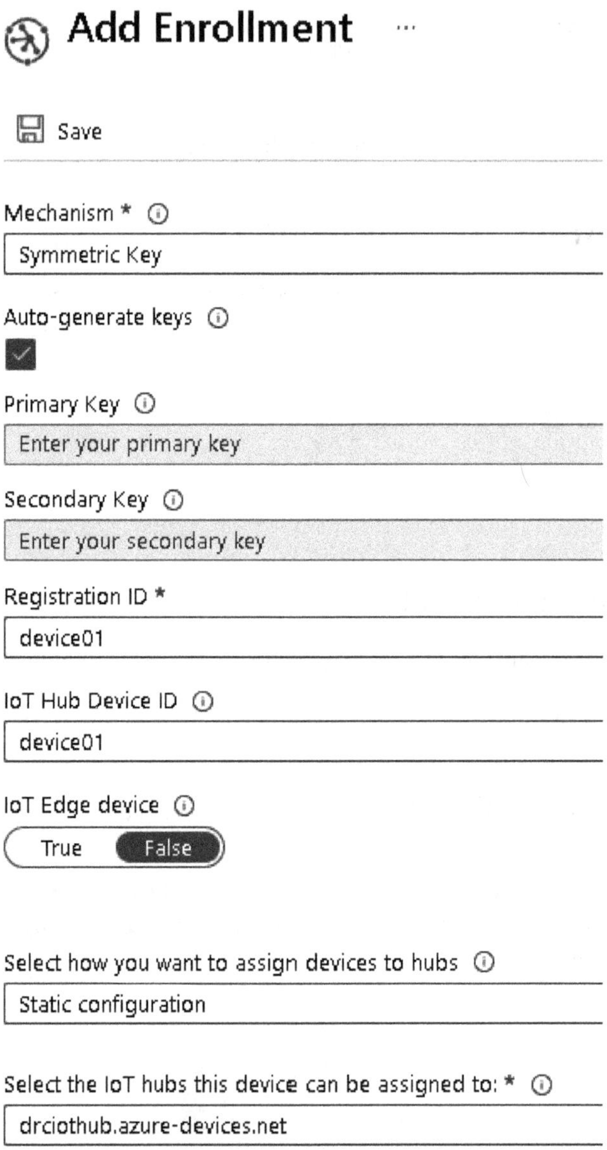

Add Enrollment   ...

🖫 Save

Mechanism * ⓘ

| Symmetric Key |

Auto-generate keys ⓘ

☑

Primary Key ⓘ

| Enter your primary key |

Secondary Key ⓘ

| Enter your secondary key |

Registration ID *

| device01 |

IoT Hub Device ID ⓘ

| device01 |

IoT Edge device ⓘ

(True   **False**)

Select how you want to assign devices to hubs ⓘ

| Static configuration |

Select the IoT hubs this device can be assigned to: * ⓘ

| drciothub.azure-devices.net |

Figure 3.6 – Adding a device enrollment

5. Once you've saved the individual enrollment, open it back up by clicking on it in the enrollment list.

6. Notice that the primary and secondary keys have been created. You will need to use these later.

7. In this book's GitHub repository, open the starter folder for lab 3.1 in VS Code. If VS Code asks you to load the required assets, select yes.

8. Open the Program.cs file in the VS Code and replace the dpsIdScope variable with the DPS service's **ID Scope**; this can be found on the **Overview** page of the DPS service in the Azure portal:

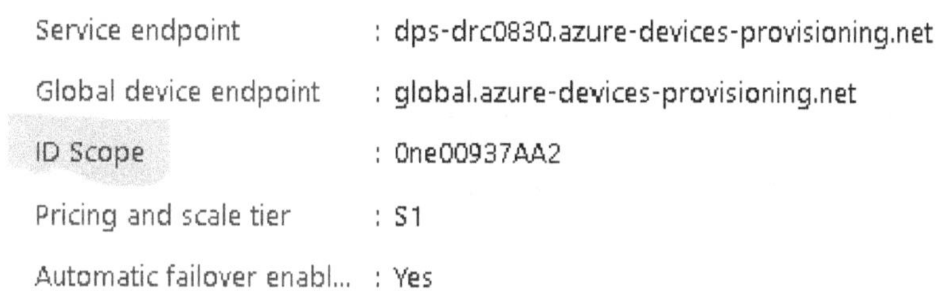

Service endpoint	: dps-drc0830.azure-devices-provisioning.net
Global device endpoint	: global.azure-devices-provisioning.net
ID Scope	: 0ne00937AA2
Pricing and scale tier	: S1
Automatic failover enabl...	: Yes

Figure 3.7 – Getting the ID Scope from the DPS service

9. Set device01 as the registrationId variable.

10. Update the individualEnrollmentPrimaryKey and individualEnrollmentSecondaryKey variables using the **Primary Key** and **Secondary Key** values that were listed for device01 in the DPS service.

11. To create the Main method of your simulated device, enter the following code where it says // INSERT Main method below here (this code snippet is available in the *Chapter 3* starter file in this book's GitHub repository):

```
public static async Task Main(string[] args)
    {
```

12. First, set up the keys:

```
        using (var security = new
SecurityProviderSymmetricKey(registrationId,
                        individualEnrollmentPrimaryKey,
                        individualEnrollmentSecondaryKey))
```

13. Next, set up the transport type and the device client:

```
            using (var transport = new
ProvisioningTransportHandlerAmqp(TransportFallbackType.TcpOnly))
        {
            ProvisioningDeviceClient provClient =
ProvisioningDeviceClient.Create(GlobalDeviceEndpoint,
dpsIdScope, security, transport);

            using (deviceClient = await
ProvisionDevice(provClient, security))
        {
            await deviceClient.OpenAsync().
ConfigureAwait(false);
```

14. Start reading and sending device telemetry:

```
            Console.WriteLine("Start reading and sending
device telemetry...");
            await
SendDeviceToCloudMessagesAsync(deviceClient);

            await deviceClient.CloseAsync().
ConfigureAwait(false);
        }
    }
}
```

15. To create the `ProvisionDevice` method of your simulated device, enter the following code where it says `// INSERT ProvisionDevice method below here`:

```
private static async Task<DeviceClient>
ProvisionDevice(ProvisioningDeviceClient
provisioningDeviceClient, SecurityProviderSymmetricKey security)
    {
        var result = await provisioningDeviceClient.
RegisterAsync().ConfigureAwait(false);
        Console.WriteLine($"ProvisioningClient AssignedHub:
{result.AssignedHub}; DeviceID: {result.DeviceId}");
        if (result.Status !=
ProvisioningRegistrationStatusType.Assigned)
        {
            throw new Exception($"DeviceRegistrationResult.
Status is NOT 'Assigned'");
        }
        var auth = new
DeviceAuthenticationWithRegistrySymmetricKey(
            result.DeviceId,
```

```
                    security.GetPrimaryKey());

                return DeviceClient.Create(result.AssignedHub, auth,
        TransportType.Amqp);
            }
```

16. Navigate to **View** | **Terminal**. In the **Terminal** pane, ensure the command prompt shows the directory path for the `Program.cs` file.

17. To build and run the **Simulated Device** application, enter the `dotnet run` command..

18. As soon as the **Simulated Device** application is launched, it will start transmitting telemetry events to Azure IoT Hub. These events contain various data points, such as temperature, humidity, pressure, latitude, and longitude. Additionally, the same information will be displayed on the terminal screen.

19. Back in the Azure portal, open **Bash** in Cloud Shell. Enter the following CLI command, replacing `{IOTHubname}` with your IoT Hub's name:

```
az iot hub monitorevents --hub-name {IoTHubName} --device-id
device01
```

20. Verify that the hub is receiving sensor readings.

21. Use *Ctrl* + *C* to stop both the `az` command and the **Simulated Device** application.

Adding group enrollments

The following steps describe how we can add group enrollments in DPS:

1. Open your DPS service in the Azure portal, go to the **Manage enrollments** tab and select **Add a group enrollment**.

2. Add a group named `group01` with auto-generated keys and static configuration that's linked to your IoT Hub:

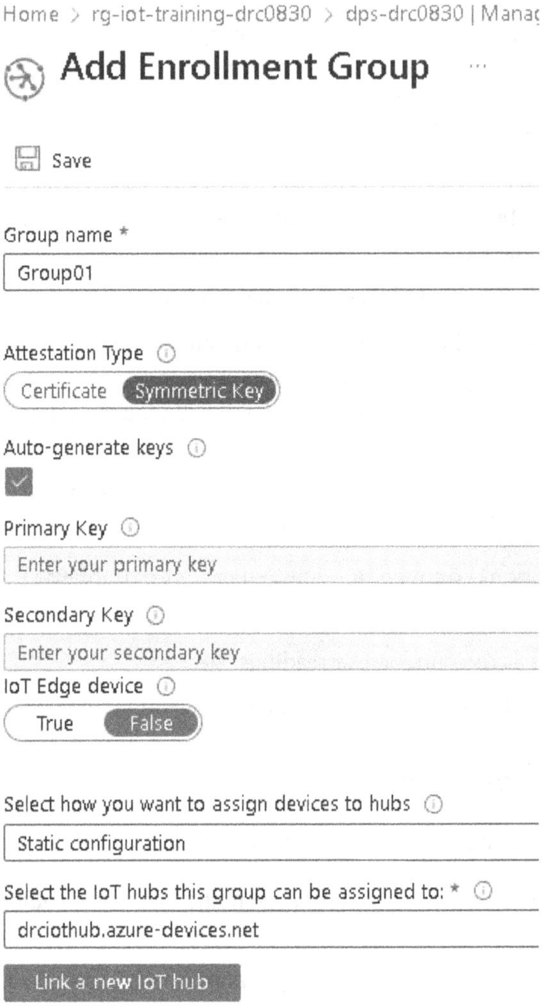

Home > rg-iot-training-drc0830 > dps-drc0830 | Manag

Add Enrollment Group   ...

💾 Save

Group name *

Group01

Attestation Type ⓘ

(Certificate **Symmetric Key**)

Auto-generate keys ⓘ

✓

Primary Key ⓘ

Enter your primary key

Secondary Key ⓘ

Enter your secondary key

IoT Edge device ⓘ

(True **False**)

Select how you want to assign devices to hubs ⓘ

Static configuration

Select the IoT hubs this group can be assigned to: * ⓘ

drciothub.azure-devices.net

Link a new IoT hub

Figure 3.8 – Adding an enrollment group

3. After saving the group, re-open it and copy its **Primary Key**.

4. Generate a device key by using the following command in the **Bash** terminal. Replace the value
    of the `--key` parameter with the **Primary Key** value from your enrollment group:

    ```
    az iot dps enrollment-group compute-device-key --key <group
    primary key> --registration-id device1.1
    ```

5. Copy the derived key that's returned. This is the group device key for the device.

6. Open the `azure-iot-sdk-python\samples\async-hub-scenarios` folder in
    VS Code.

7. You can look at the file in the editor and observe the code.

8. In the command prompt, switch to the directory that contains the `provision_symmetric_` `key_group.py` file. Run the following code to set the environment variables that will be used by the code, replacing `<id-scope>`, `<registration-id>`, and `<derived-device-key>` with your unique values:

```
set PROVISIONING_HOST=global.azure-devices-provisioning.net
set PROVISIONING_IDSCOPE=<id-scope>
set PROVISIONING_REGISTRATION_ID=<registration-id>
set PROVISIONING_SYMMETRIC_KEY=<derived-device-key>
```

9. If you have not already installed it, install the `azure-iot-device` library by running the following command:

```
pip install azure-iot-device
```

10. Run the following code:

```
python provision_symmetric_key.py
```

11. You should see the output of the device sending messages to the IoT device:

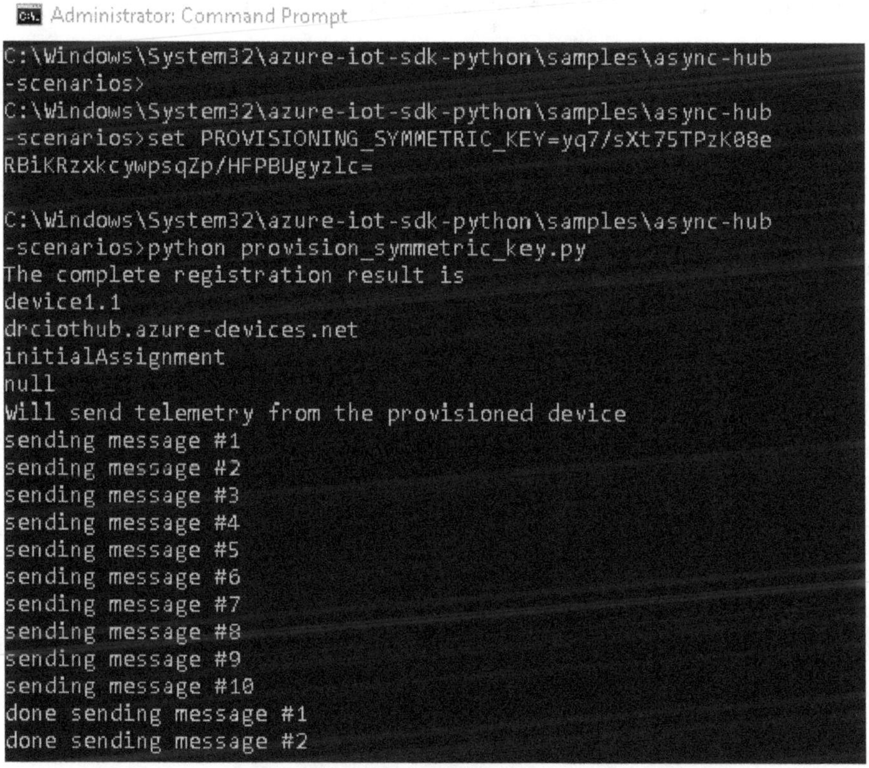

Figure 3.9 – Device sending messages

12. Verify that the device has been registered under the group in the Azure portal in the DPS service.

Now that you know how to work with DPS, IoT Hub, and devices to create enrollments, let's see how we can customize the allocation of devices.

Using custom policies for device allocation

In the context of IoT, it is crucial to have an efficient allocation policy to manage IoT devices effectively. The allocation policy determines how the IoT devices are assigned to IoT Hubs, which are responsible for managing and processing the data that's collected from the devices. In this lab, we will do the following:

- Create a custom allocation policy
- Test the allocation policy

To create a custom policy, complete the following steps:

1. In Azure, create a resource group called `IoTLab-rg`.
2. In the resource group, create two standard IoT Hubs named `iotlabhub01` and `iotlabhub02`.
3. In the same resource group, create a DPS service named `dps01`.
4. Link your IoT Hubs to the DPS service.
5. To create the custom allocation function, in the same resource group, select **Create** at the top of the page.
6. Search for and create a Function App instance.
7. Name the function app `customAllocation` with a random set of numbers to make it unique.
8. Under **Publish**, select **Code**. The runtime stack is .NET version 6 (LTS).
9. Set **Operating system** to **Windows** and **Plan type** to **Consumption (Serverless)**.
10. Under the **Monitoring** tab, select **No** for **Enable Application insights**.
11. Create the function app.
12. Once created, go to the function app. From the left-hand side menu, select **Functions**. At the top of the page, select **Create**.
13. When the **Create Function** page opens, select **Develop in portal** in the development environment and the **HTTP trigger** template.
14. Name the function `funcCustAllow` and set **Authorization level** to `Function`.
15. Create the function.
16. Once the function has been created, select the **Code + Test** option from the left-hand side menu.

17. Replace the code in the `run.csx` file with the following code (you can find this code in the GitHub starter folder for *Chapter 3*):

I.   First, add your `using` statements to reference the APIs you will be using:

```
#r "Newtonsoft.Json"

using System.Net;
using Microsoft.AspNetCore.Mvc;
using Microsoft.Extensions.Primitives;
using Newtonsoft.Json;
```

II.   Next, define the task:

```
public static async Task<IActionResult> Run(HttpRequest req,
ILogger log)
{
    log.LogInformation("C# HTTP trigger function processed a
    request.");
```

III.   Get the request body:

```
string requestBody = await new StreamReader(req.Body).
ReadToEndAsync();
dynamic data = JsonConvert.DeserializeObject(requestBody);

log.LogInformation("Request.Body:...");
log.LogInformation(requestBody);

// Get registration ID of the device
string regId = data?.deviceRuntimeContext?.registrationId;
```

IV.   Trap for errors:

```
string message = "Uncaught error";
bool fail = false;
ResponseObj obj = new ResponseObj();

if (regId == null)
{
    message = "Registration ID not provided for the device.";
    log.LogInformation("Registration ID : NULL");
    fail = true;
}
else
```

```
    {
       string[] hubs = data?.linkedHubs?.ToObject<string[]>();
```

V.   Verify the hub:

```
// Must have hubs selected on the enrollment
if (hubs == null)
{
  message = "No hub group defined for the enrollment.";
  log.LogInformation("linkedHubs : NULL");
  fail = true;
}
else
{
  if (regId.Contains("hub01"))
  {
    foreach(string hubString in hubs)
    {
      if (hubString.Contains("hub01"))
      obj.iotHubHostName = hubString;
    }

    if (obj.iotHubHostName == null)
    {
      message = "No hub found for the enrollment.";
      log.LogInformation(message);
      fail = true;
    }
    else
    {
    }
  }
  else if (regId.Contains("hub02"))
  {
    //Find the IoT hub configured on the enrollment
    foreach(string hubString in hubs)
    {
      if (hubString.Contains("hub02"))
      obj.iotHubHostName = hubString;
    }
    if (obj.iotHubHostName == null)
    {
      message = "No hub found for the enrollment.";
      log.LogInformation(message);
```

```
        fail = true;
    }
    else
    {
    }
}
```

VI.   Check for unrecognized devices:

```
    // Unrecognized device.
    else
    {
        fail = true;
        message = "Unrecognized device registration.";
        log.LogInformation("Unknown device registration");
    }
    }
}

    Log the information.log.LogInformation("\nResponse");
    log.LogInformation((obj.iotHubHostName != null) ? JsonConvert.
    SerializeObject(obj) : message);
    return (fail)
    ? new BadRequestObjectResult(message)
    : (ActionResult)new OkObjectResult(obj);
}
public class ResponseObj
{
    public string iotHubHostName {get; set;}
}
```

18. To use the allocation function, in the DPS service, click **Manage Enrollments** under **Settings** from the left-hand side menu.

19. At the top of the page, select **Add enrollment group**.

20. Name the group `allocationTest` with **Attestation Type** set to **Symmetric Key**.

21. Select **Custom** for assigning devices to hubs. Then, select the function we created in the previous section and save the group.

22. To enroll a device using the group, we must create a device key from the group key. Run the following code in the Bash portal for `device01hub01` and again for `device02hub02`. Replace the key with your group key. Save the keys that are generated for later:

```
az iot dps enrollment-group compute-device-key --key
G8oMdBndzeDdsdoDdlCZ4g9vg5XePE+KtelF1pyGnszw8RKqytYkQnwx
VUwwMiVZ22iccNoNIBUnYKubgKCbeg== --registration-id device01hub01
```

23. To test the group enrollment, you will create a simulated device that calls the DPS service to be assigned to the correct IoT Hub based on the function we created earlier.

24. Open the `Lab3.2 Device Simulator` folder in VS Code. Review the code in the `ProvisioningDeviceClientSample.cs` and `Parameters.cs` files. Review the code for provisioning the device. When you run the code, you will need to provide a parameter for `Id-scope`, `registation-id`, and the primary key.

25. To get the DPS service's `id-scope`, go to **Overview** in the left menu of the DPS service. You should see **ID Scope** listed, as shown in the following figure:

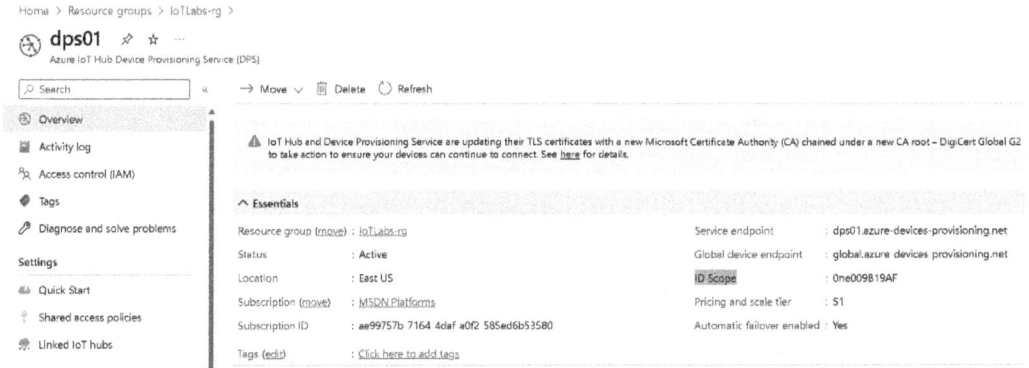

Figure 3.10 – Getting the ID scope

26. You generated the primary key using the group key in *step 5*. The registration ID for the device will be `device01hub01` or `device02hub02`, depending on which device you are simulating.

27. In the Windows command prompt, navigate to the `simulated devices` folder and run the following code. Replace the parameters with your values:

```
dotnet run --i <id-scope> --r <registration-id> --p <primarykey>
```

28. You should see an output message confirming its registration and the hub it was assigned to:

```
Device device01hub01 registered to iotlabhub01.azure-devices.
net.
```

29. Test the other device and then clean up your resources.

Now, let's take a look at what we learned in this chapter and where we are heading next.

Summary

In this chapter, you explored the features of DPS. Although we only used a few devices in the examples provided, DPS allows you to provision millions of devices in a secure and scalable manner.

You learned some important concepts alongside device provisioning and saw how security is an integral part of the process. You have the option to use keys, certificates, or TPM. You also saw how allocation policies work and how to create custom allocation policies through functions. Last but not least, you learned how to implement group enrollments.

Now that you can effectively deploy thousands of devices, you need to learn how to monitor and manage them. In *Chapter 4, Device Management and Monitoring*, you will learn about what tools and processes you can use to monitor and manage your devices at scale.

4

Exploring Device Management and Monitoring

This chapter looks at **device management** and how Azure IoT Hub automates many of the repetitive and complex tasks of managing large sets of devices. You will investigate how **device twins** are an essential tool for automating device management. You will also discover the essential monitoring and logging tools required for a secure IoT system.

In this chapter, we will cover the following main topics:

- Introducing device twins
- Communicating with devices
- Automated device management
- Monitoring metrics and logs
- Lab – Automating IoT device management
- Lab – Creating and testing an alert

Introducing device twins

Device twins are a way of storing state data of your devices. They are JSON documents that contain desired properties, reported properties, and tags. Device twins are extensively used in IoT to store device metadata in the cloud, report the current state of your devices, synchronize the state of long-running processes, and query device metadata.

For example, you can use a device twin to store configuration settings for each device. This can include things such as sampling intervals, alert thresholds, and communication settings—for instance:

Sampling interval: 10 minutes

High temperature alert threshold: 30°C

Low humidity alert threshold: 20%

The following screenshot shows a sample device twin for a device:

```json
{
    "deviceId": "device01",
    "etag": "AAABBBBBc=",
    "status": "enabled",
    "statusReason": "provisioned",
    "statusUpdateTime": "0001-01-01T00:00:00",
    "connectionState": "connected",
    "lastActivityTime": "2015-02-30T16:24:48.789Z",
    "cloudToDeviceMessageCount": 0,
    "authenticationType": "sas",
    "x509Thumbprint": {
        "primaryThumbprint": null,
        "secondaryThumbprint": null
    },
    "version": 2,
    "tags": {
        "deploymentLocation": {
            "building": "12",
            "floor": "5"
        }
    },
    "properties": {
        "desired": {
            "telemetryConfig": {
                "sendFrequency": "10m"
            },
            "$metadata" : {...},
            "$version": 1
        },
        "reported": {
            "telemetryConfig": {
                "sendFrequency": "10m",
                "status": "success"
            },
            "batteryLevel": 75,
            "$metadata" : {...},
            "$version": 4
        }
    }
}
```

Figure 4.1: Sample device twin

The preceding code is also available on the GitHub repo in the `Chapter04` folder.

Tags are set by a backend application or through IoT Hub. They are used to organize and query your devices. The device itself does not read tags or write to tags. **Desired properties** are also set by a backend application or through IoT Hub. They can be read by the device. A **reported property** is set by the device and is read either by a backend application or through IoT Hub. There are also some metadata and versioning tags.

Tags and properties support optimistic concurrency and have an `etag` property that can be used for conditional updates. There is also a `$version` tag that is incremental and can be used to enforce consistency of updates.

Now that we know what device twins are, let's look at how they can be used.

Querying devices using device twins

You often need to query for a set of devices either to run a job, route messages, or create alerts. Luckily, IoT Hub provides a robust SQL-like query language. You can use this directly in IoT Hub or in an application using the appropriate SDKs.

The following screenshot shows the use of the query editor in IoT Hub. The queries use the familiar `SELECT` and `FROM` clauses. You can also use `WHERE` and `GROUP BY` clauses:

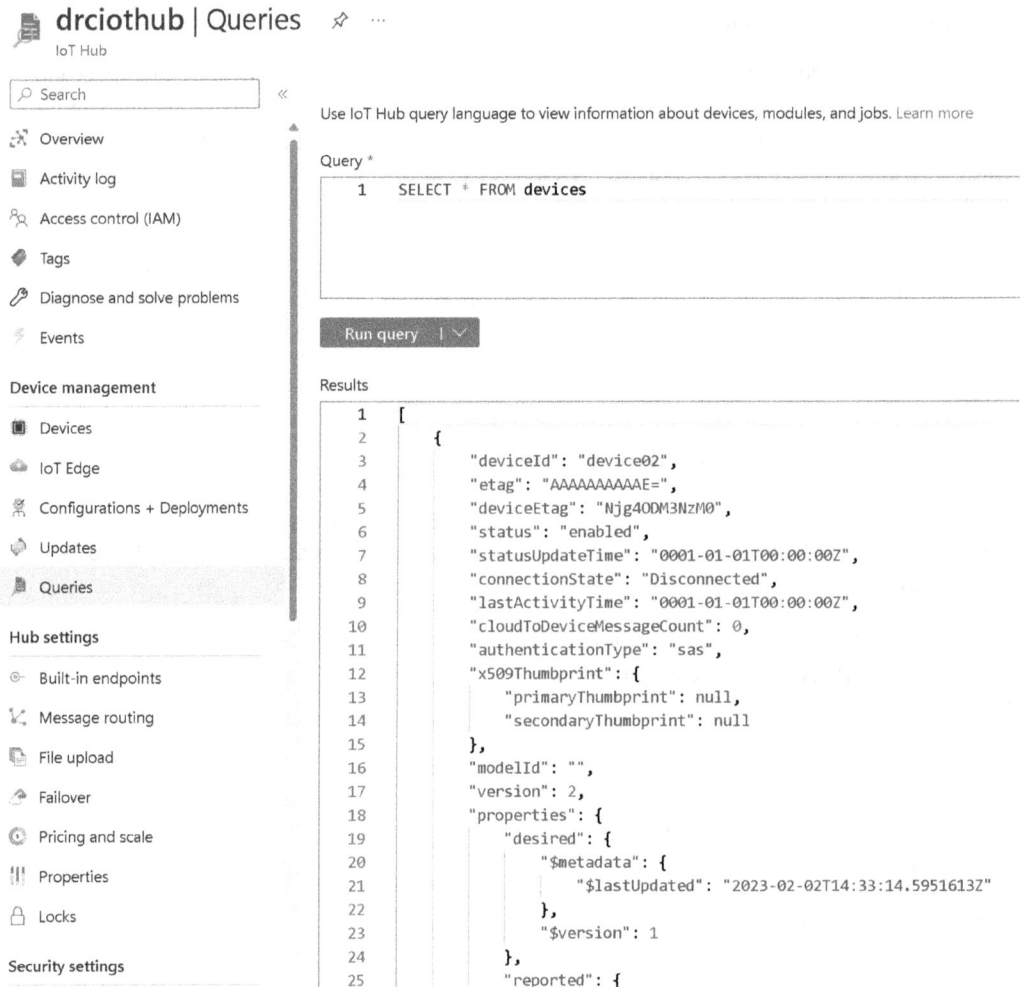

Figure 4.2: The query editor in IoT Hub

Notice that that the query returns full JSON twin documents. This is because we used SELECT *. You can narrow down the results by selecting specific tags. The following screenshot shows a query that pulls information from specific tags. Notice that that you can easily walk the document for nested tags:

Query *

```
1   SELECT  deviceId, properties.reported.$version, properties.reported.$metadata.$lastUpdated
2   FROM devices
```

Run query | ∨

Results

```
1   [
2       {
3           "deviceId": "device02",
4           "$version": 1,
5           "$lastUpdated": "2023-02-02T14:33:14.5951613Z"
6       },
7       {
8           "deviceId": "device1.1",
9           "$version": 1,
10          "$lastUpdated": "2023-02-05T19:49:18.2299519Z"
11      },
12      {
13          "deviceId": "device01",
14          "$version": 1,
15          "$lastUpdated": "2022-12-29T18:11:00.2507379Z"
16      }
17  ]
```

Figure 4.3: Querying information from specific tags

You can also query device twins from your own applications using the device SDK. The following C# code snippet queries for devices in HBG and using a cellular network:

```
var query = registryManager.CreateQuery(
    "SELECT * FROM devices WHERE tags.location.plant = 'HBG'
        AND properties.reported.connectivity.type = 'cellular'", 100);
var twinsInHBGUsingCellular = await query.GetNextAsTwinAsync();
Console.WriteLine("Devices in HBG using cellular network: {0}",
string.Join(", ", twinsInHBGUsingCellular.Select(t => t.DeviceId)));
```

Updating device properties

To synchronize state across devices, you can use desired properties and the the reported properties of the device twin. When a device connects to IoT Hub, it reads the device twin and synchronizes its reported properties and desired properties. For example, if your device is controlling the pressure of a steam valve, it compares the desired pressure with its reported pressure and adjusts the pressure accordingly. Devices can also be notified when the desired pressure is updated.

You can create an event handler on a device that responds to updates to the device twin. A device has a local twin property and compares it to the delta sent from the backend. The following JavaScript code handles a change in the pressure node of the device twin:

```
// Handle desired properties updates to the pressure component
twin.on('properties.desired.components.pressure', function(delta) {
    if (delta.minPressure || delta.maxpressure) {
        console.log(chalk.green('\nUpdating desired pressure in pressure
component:'));
        console.log('Updating minimum pressure: ' + twin.properties.
desired.components.pressure.minPressure);
        console.log('Updating maximum pressure: ' + twin.properties.
desired.components.pressure.maxPressure);
        // Call code to reset the pressure
        // Update the reported properties and send them to the hub
        reportedPropertiesPatch.minPressure = twin.properties.desired.
components.climate.minPressure;
        reportedPropertiesPatch.maxPressure = twin.properties.desired.
components.climate.maxPressure;
        sendReportedProperties();
    }
});
```

Along with device twin desired properties, there are two other ways to communicate with devices: direct methods and cloud-to-device messaging. We will discuss these in the following section.

Communicating with devices

Direct methods are for use when you need a direct response back, while cloud-to-device messaging is one-way communication to a device. Let us look a little deeper into these two communication methods.

Cloud-to-device messaging

Cloud-to-device messages are one-way communications to a device. They are sent to a queue that guarantees *at-least-once delivery* and are best used when there is intermittent connectivity of devices.

Although you can send messages to your devices in the Azure portal, you are more likely to send messages from a backend application using SDKs. The following C# code demonstrates the device code responding to a cloud-to-device message:

```
private static async void ReceiveMessage()
{
    Console.WriteLine("\nReceiving messages.");
    while (true)
    {
```

```
        Message receivedMessage = await deviceClient.ReceiveAsync();
        if (receivedMessage == null) continue;

            Console.WriteLine("Received message: {0}",
        Encoding.ASCII.GetString(receivedMessage.GetBytes()));

        await deviceClient.CompleteAsync(receivedMessage);
    }
}
```

The device notifies IoT Hub that the message has been received and processed using the call to `CompleteAsync()`. This indicates that the message can be safely removed from the queue.

Direct method calls

Another way to issue calls to devices is **direct methods**. They are for use when you need a direct response. With direct methods, devices have to be online to receive and respond to messages. Remember that you issue direct methods when you need an immediate response from your device. Direct methods follow the familiar request-response pattern, and devices need to be connected at the time of the call. You can invoke a method from a backend application through the REST API or using the appropriate IoT Hub service SDK. The following code shows how you would make a direct method call to a device using the service SDK:

```
private static async Task InvokeMethodAsync(string deviceId,
ServiceClient serviceClient),
{
    var methodInvocation = new
    CloudToDeviceMethod("SetTelemetryInterval")
    {
        ResponseTimeout = TimeSpan.FromSeconds(30),
    };
    methodInvocation.SetPayloadJson("10");

    Console.WriteLine($"Invoking direct method for device: {deviceId}");

    // Invoke the direct method asynchronously and get the response from
    the simulated device.
    CloudToDeviceMethodResult response = await serviceClient.
    InvokeDeviceMethodAsync(deviceId, methodInvocation);

    Console.WriteLine($"Response status: {response.Status}, payload:\n\
    t{response.GetPayloadAsJson()}");
}
```

The following code C# code demonstrates how a device responds to a direct method call. First, you set up a call handler:

```
await s_deviceClient.SetMethodHandlerAsync("SetTelemetryInterval",
SetTelemetryInterval, null);
And then the method.
private static Task<MethodResponse> SetTelemetryInterval(MethodRequest
methodRequest, object userContext)
{
  string data = Encoding.UTF8.GetString(methodRequest.Data);

  // Check the payload is a single integer value.
  if (int.TryParse(data, out int telemetryIntervalInSeconds))
  {
    s_telemetryInterval = TimeSpan.
    FromSeconds(telemetryIntervalInSeconds);
    Console.ForegroundColor = ConsoleColor.Green;
    Console.WriteLine($"Telemetry interval set to
    {s_telemetryInterval}");
    Console.ResetColor();

    // Acknowledge the direct method call with a 200 success message.
    string result = $"{{\"result\":\"Executed direct method:
    {methodRequest.Name}\"}}";
    return Task.FromResult(new MethodResponse(Encoding.UTF8.
    GetBytes(result), 200));
  }
  else
  {
    // Acknowledge the direct method call with a 400 error message.
    string result = "{\"result\":\"Invalid parameter\"}";
    return Task.FromResult(new MethodResponse(Encoding.UTF8.
    GetBytes(result), 400));
  }
}
```

As you can see, the method sends back a response of success or failure to the requestor, along with a JSON payload if necessary.

Now that you have seen how to communicate with devices, the next thing to look at is how we can automate some of the repetitive tasks required to maintain your device fleet.

Automated device management

There are many repetitive tasks when managing IoT devices. Luckily, Azure IoT provides many ways to automate these tasks. You have already seen how the **Device Provisioning Service (DPS)** is used to automate device provisioning and register devices at scale. In addition, Azure IoT Hub provides a central location where you can manage your IoT devices and data. It provides features such as message routing, telemetry ingestion, and device-to-cloud communication. It also enables you to manage devices, monitor device health, and update device firmware and software **over-the-air (OTA)**.

An **Azure IoT Hub job** is a background task that you can use to manage devices and perform bulk operations on them. These jobs are used to automate tasks that would otherwise be time-consuming or impossible to perform manually. There are several types of jobs that you can run in Azure IoT Hub, including:

- **Device configuration**: This type of job allows you to deploy a new configuration to a set of devices or update an existing configuration. For example, you might want to change a device's sampling rate or update its telemetry settings.

- **Firmware update**: This type of job allows you to update the firmware on a set of devices. This is useful when you need to fix bugs or security vulnerabilities in the device firmware.

- **Device twin update**: This type of job allows you to update the device twin properties of a set of devices. A device twin is a JSON document that contains metadata about a device, such as its capabilities, state, and configuration.

- **Export**: This type of job allows you to export device data to an external system for further analysis or processing. You can export data to Azure Blob storage, Azure Event Hubs, or a custom endpoint.

You can schedule jobs to run immediately or at a specific time in the future. You can also monitor the status of a job, cancel a running job, or rerun a completed job. Jobs can be created and managed programmatically using the Azure IoT Hub REST API or the Azure IoT Hub SDKs for various programming languages.

Setting up a job in Azure IoT Hub is a straightforward process that involves selecting the target devices, specifying the job type and configuration details, and monitoring the job's progress.

Automated device management requires the use of a configuration file in the form of a JSON document. It contains a target condition that uses a query on twin and/or reported properties and defines the scope of updates. The target content defines an update to the device twin. The metrics define a summary of the job run, such as devices targeted and the number of successful updates.

To set up a job, go to **IoT Hub** in the portal and select the **Configurations + Deployments** blade. Select **Add | Device Twin Configuration**:

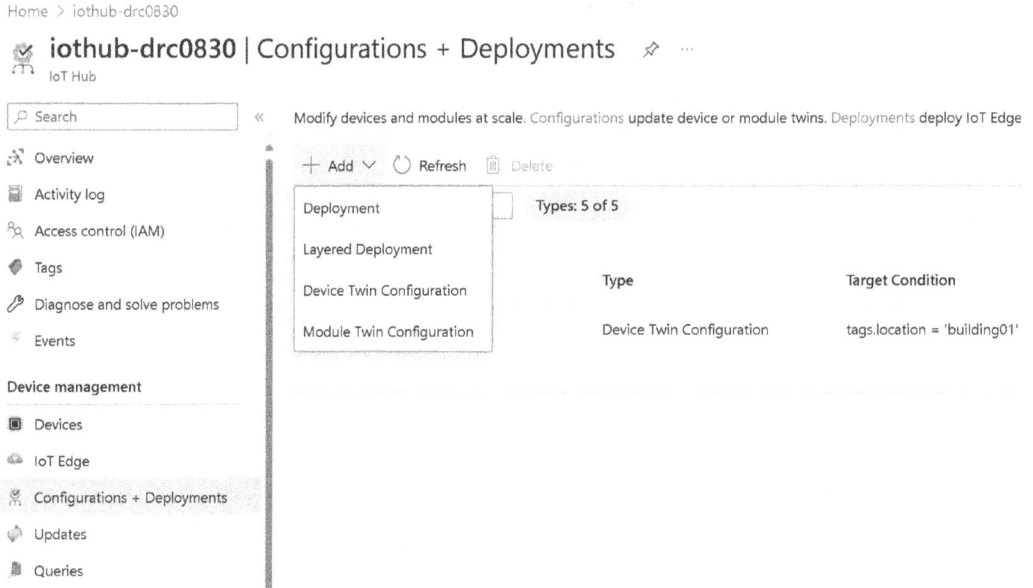

Figure 4.4: Setting up a job

This will bring up a series of screens that allow you to manage the configuration, entering **ID**, **Labels**, **Device Twin Settings**, **Target Devices** values, and any custom **Metrics** values you may need. Once saved, you can open the configuration details and update some of the settings, such as **Target Condition** and **Priority**:

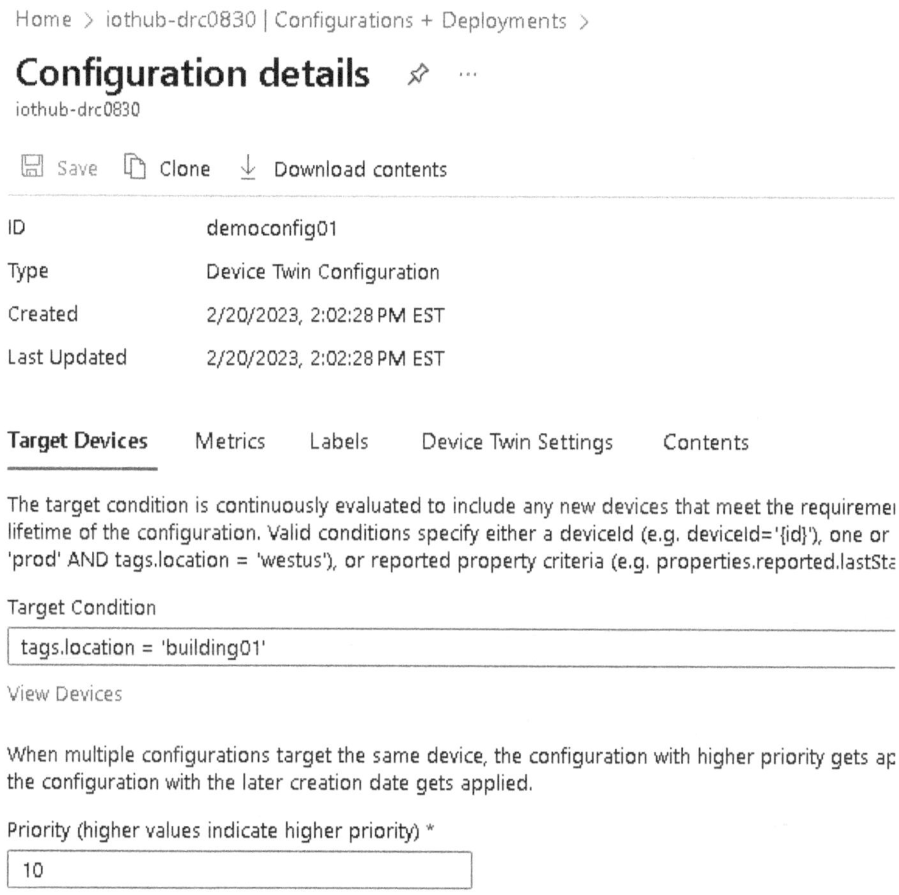

Home > iothub-drc0830 | Configurations + Deployments >

Configuration details 📌 ⋯
iothub-drc0830

💾 Save   📋 Clone   ↓ Download contents

ID	democonfig01
Type	Device Twin Configuration
Created	2/20/2023, 2:02:28 PM EST
Last Updated	2/20/2023, 2:02:28 PM EST

Target Devices   Metrics   Labels   Device Twin Settings   Contents

The target condition is continuously evaluated to include any new devices that meet the requiremei
lifetime of the configuration. Valid conditions specify either a deviceId (e.g. deviceId='{id}'), one or
'prod' AND tags.location = 'westus'), or reported property criteria (e.g. properties.reported.lastSta

Target Condition

    tags.location = 'building01'

View Devices

When multiple configurations target the same device, the configuration with higher priority gets ap
the configuration with the later creation date gets applied.

Priority (higher values indicate higher priority) *

    10

Figure 4.5: Configuring a job

You can also view the JSON document created for you:

Configuration

The text box below displays the configuration to be submitted.

```
1    {
2        "content": {
3            "deviceContent": {
4                "properties.desired.room-temp": {
5                    "Temp": 66
6                }
7            }
8        },
9        "etag": "",
10       "labels": {
11           "Version": "1"
12       },
13       "metrics": {
14           "queries": {}
15       },
16       "id": "democonfig01",
17       "priority": 10,
18       "targetCondition": "tags.location = 'building01'"
19   }
```

Figure 4.6: Reviewing the JSON document

You can schedule jobs using any of the supporting languages. The following C# code uses the Azure IoT service SDK and the `JobClient` class:

```
public static async Task StartMethodJob(string jobId)
{
    CloudToDeviceMethod directMethod =
        new CloudToDeviceMethod("TurnLightOff", TimeSpan.
FromSeconds(10),
        TimeSpan.FromSeconds(10));

    JobResponse result = await jobClient.
ScheduleDeviceMethodAsync(jobId,
        $"DeviceId IN ['{deviceId}']",
        directMethod,
        DateTime.UtcNow,
        (long)TimeSpan.FromMinutes(5).TotalSeconds);

    Console.WriteLine("Started Method Job");
}
```

Automation through jobs is an important aspect of large fleet management of devices, increasing both efficiency and productivity. Another crucial aspect of fleet management is monitoring devices for unusual behavior. Let's look at this next.

Monitoring metrics and logs

As with most services in Azure, IoT Hub uses **Azure Monitor**. Azure Monitor is a monitoring service provided by Microsoft Azure that enables you to collect and analyze telemetry data from your Azure resources and applications. It allows you to monitor the performance and availability of your applications, infrastructure, and network resources in real time and provides insights and alerts to help you troubleshoot issues and optimize performance.

Most services have a set of metrics used for monitoring and alerting that are tailored to them. For IoT Hub, these include:

- Cloud-to-device command metrics
- Cloud-to-device direct method metrics
- Cloud-to-device twin operations metrics
- Configurations metrics
- Daily quota metrics
- Device metrics
- Device telemetry metrics
- Device-to-cloud twin operations metrics
- Event Grid metrics
- Jobs metrics
- Routing metrics
- Twin query metrics

As you can see, there are quite a lot of metrics you need to monitor. For example, using IoT cloud-to-device command metrics, you can remotely monitor and control your equipment, based on real-time sensor data. This can help you to optimize the performance of your equipment, reduce downtime, and improve overall efficiency. Using IoT cloud-to-device direct methods, you can remotely manage and update your IoT devices, without requiring any action from the device or user. This can help you to improve the security, reliability, and functionality of your IoT devices. IoT cloud-to-device twin operations allow you to maintain a digital twin of each of your IoT devices in the cloud. A digital twin is a virtual representation of a physical device that reflects its current state, configuration settings, and metadata. You can use the twin to remotely monitor and manage the state of your devices and ensure that they are consistent across all devices.

To set up your metrics, in the **Monitoring** section of the **IoT Hub** blade, select **Metrics**. This will present a graph where you can select different metrics to display:

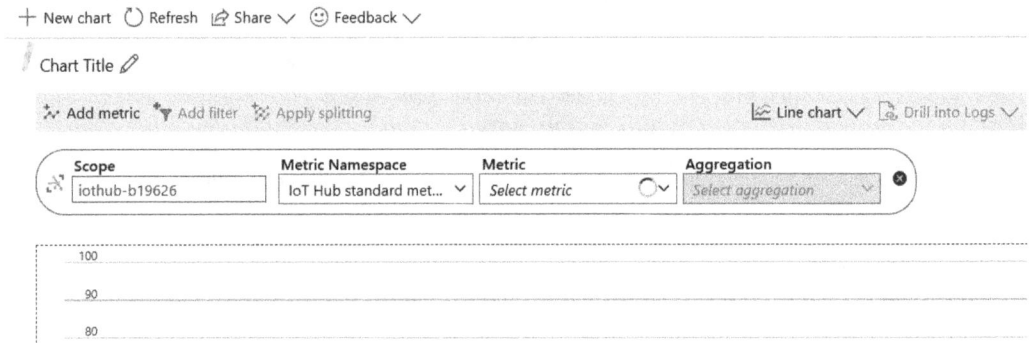

Figure 4.7: Selecting metrics

You can add several metrics to a graph, select different types of graphs, perform various aggregations, create alerts, and pin a graph to a dashboard. If you select **Logs** in the **Monitoring** section of the **IoT Hub** blade, you get a screen to issue queries against the logs. There are sample queries you can choose from, and you can also write your own. The queries are written in **Kusto Query Language** (**KQL**). It is similar to SQL and should be easy to pick up. Before you can query the log files, you have to set up what gets logged in the **Diagnostic Settings** blade. The following KQL queries can be used for devices that have a connection error:

```
AzureDiagnostics
| where ResourceProvider == "MICROSOFT.DEVICES" and ResourceType ==
"IOTHUBS"
| where Category == "Connections" and Level == "Error"
```

You can create alerts from either the **Metrics** or the **Logs** query page. You can also set one up using the **Alerts** page under the **Monitoring** group on the left menu of the **IoT Hub** blade. The following screenshot shows how to set up an alert rule in the portal:

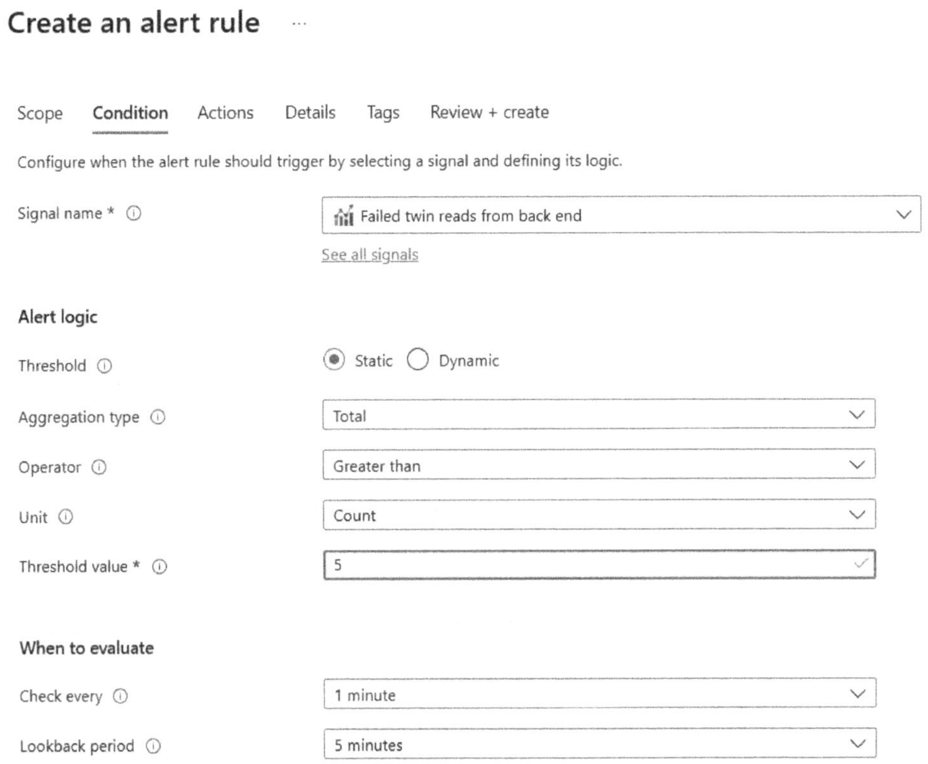

Create an alert rule ⋯

Scope   **Condition**   Actions   Details   Tags   Review + create

Configure when the alert rule should trigger by selecting a signal and defining its logic.

Signal name * ⓘ

| 📊 Failed twin reads from back end | ⌄ |

See all signals

Alert logic

Threshold ⓘ   ◉ Static   ○ Dynamic

Aggregation type ⓘ

| Total | ⌄ |

Operator ⓘ

| Greater than | ⌄ |

Unit ⓘ

| Count | ⌄ |

Threshold value * ⓘ

| 5 | ⌄ |

When to evaluate

Check every ⓘ

| 1 minute | ⌄ |

Lookback period ⓘ

| 5 minutes | ⌄ |

Figure 4.8: Creating an alert

In addition to sending out alerts, you can also automate the response. Let's take a closer look at how we can do this.

Lab – Automating IoT device management

In this exercise, you will gain an understanding of using Azure IoT Hub automatic device management. Although you will only use one device, this is typically used for many devices. To set up the environment, perform the following tasks:

1. In a browser, navigate to your Azure portal and create a resource group and an IoT hub in the resource group.

2. In the IoT hub, create a device named `device01` using symmetric key authentication.

3. Next, download the code for the simulated device from the GitHub repository's `Chapter04` starter folder.

4. Open the `simulatedDevice` folder in VS Code. Review the code in the project file. This contains the referenced NuGet packages.

5. Open the program file and find the following line of code:

    ```
    private readonly static string deviceConnectionString = "<your
    device connection string>";
    ```

6. Enter your device connection string.

7. Save and then run the app.

8. You should see the following output in the terminal:

    ```
    Device booted
    Current firmware version: 1.0.0
    ```

Now, we create the device management configuration:

1. Leave the simulated device up and running, and in the Azure portal, navigate to the **IoT Hub** page.

2. On the left-side navigation menu, under **Device management**, click **Configurations + Deployments**.

3. Choose to add a **Device Twin Configuration** type. Name the configuration `Update`.

4. Choose the **Next: Twins Settings >** option.

5. In the **Device Twin Property** field, enter the following: `properties.desired.firmware`.

6. In the **Device Twin Property Content** field, enter the following:

    ```
    {
        "fwVersion":"1.0.1",
        "fwPackageURI":"https://MyPackage.uri",
        "fwPackageCheckValue":"1234"
    }
    ```

7. Move to the **Target Devices** tab. Under **Priority**, set the value to 2. In the **Target Condition field**, enter the following query

    ```
    deviceId='<your device id>'.
    ```

8. Make sure you enter your device ID.

9. On the **Metrics** tab, under **Metric Name**, enter `fwupdated`, and under **Metric criteria**, enter the following code:

    ```
    SELECT deviceId FROM devices WHERE properties.reported.firmware.
    currentFwVersion='1.0.1'
    ```

10. Click the **Next: Review + Create** button at the bottom of the page. If the validation passes, click the **Create** button.

11. Once it is created, it will start looking for devices that need updating.

12. Switch back to your simulated device in VS code. You should see new messages confirming the update.

13. Stop the simulated device in VS Code by pressing the *Enter* key in the terminal.

14. Clean up your resources.

Lab – Creating and testing an alert

Use Azure Monitor to collect metrics and logs from your IoT hub to monitor the operation of your solution and troubleshoot issues. In this lab, you'll learn how to create charts based on metrics, how to create alerts that trigger on metrics, how to send IoT Hub operations and errors to Azure Monitor Logs, and how to check logs for errors.

You will perform the following tasks:

1. Utilize the Azure CLI to initiate the creation of an IoT hub, enroll a virtual device, and establish a Log Analytics workspace.

2. Transfer resource logs for IoT hub connections and device telemetry to Azure Monitor Logs, situated within the **Log Analytics** workspace.

3. Employ **Metric Explorer** to formulate a graph that reflects preferred metrics, and attach it to your dashboard.

4. Set up metric notifications to obtain email alerts for significant circumstances.

5. Retrieve and execute an application that simulates an IoT device sending messages to the IoT hub.

6. Monitor notifications for relevant alerting.

7. Inspect the **Metrics** chart on your dashboard.

8. Scrutinize IoT Hub operations and errors through Azure Monitor Logs.

First, set up the resources:

1. Log in to your Azure portal and launch the Bash environment.

2. Run the following CLI commands to create values for the parameters used to create resources:

```
location=westus
resourceGroup=iotlabs
iotDeviceName=iothub-Test-Device
```

3. Some names have to be unique. The following code creates the necessary unique parameters:

```
randomValue="enter a random value here"
iotHubName=labHub$randomValue
echo "IoT hub name = " $iotHubName
workspaceName=workspace$randomValue
echo "Log Analytics workspace name = " $workspaceName
```

4. Now that we have the necessary parameters, create a resource group:

```
az group create --name $resourceGroup --location $location
```

5. Next, create an IoT hub:

```
az iot hub create --name $iotHubName --resource-group
$resourceGroup --partition-count 2 --sku F1 --location $location
```

6. Create a Log Analytics workspace:

```
az monitor log-analytics workspace create --resource-group
$resourceGroup --workspace-name $workspaceName --location
$location
```

7. Create an IoT device entity:

```
az iot hub device-identity create --device-id $iotDeviceName
--hub-name $iotHubName
```

8. Retrieve and copy the primary connection string of the device. You will need it later:

```
az iot hub device-identity connection-string show --device-id
$iotDeviceName --hub-name $iotHubName
```

Next, let's collect the logs:

1. Remember that to view the logs, you must create a diagnostic setting indicating what is collected and where it is sent. You want to send them to the Log Analytics workspace you just created.

2. In the Azure portal, select the IoT hub, and then select the **Diagnostic** setting from the **Monitoring** section of the navigation menu and select the **Add Diagnostic** setting.

3. In the **Diagnostic setting** pane, choose **Connections** and **Device Telemetry**:

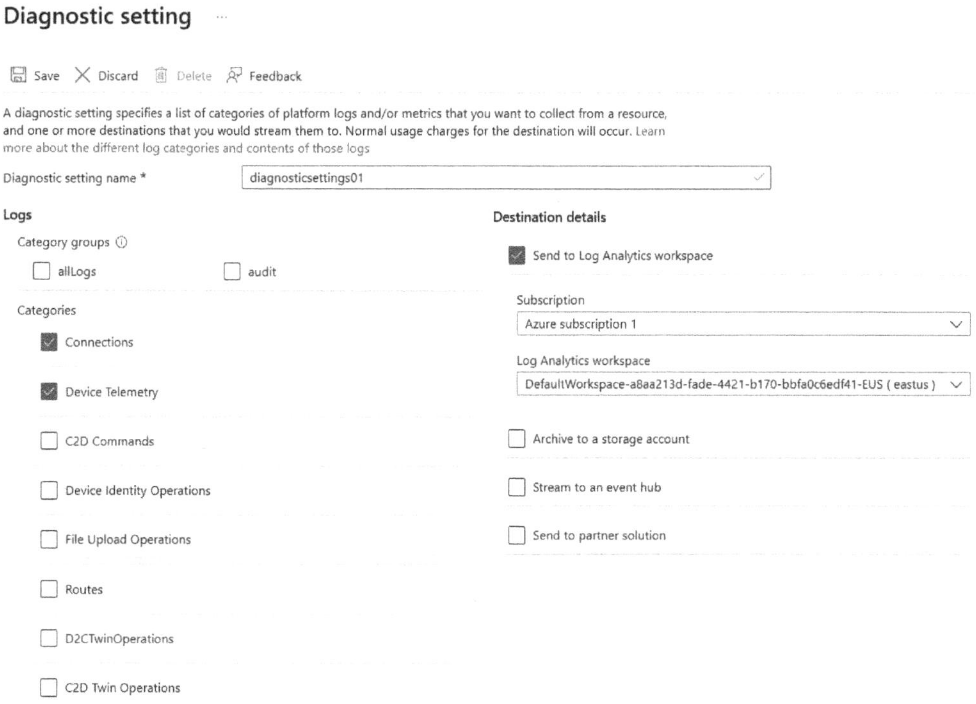

Diagnostic setting ...

🖫 Save   ✕ Discard   🗑 Delete   🗬 Feedback

A diagnostic setting specifies a list of categories of platform logs and/or metrics that you want to collect from a resource,
and one or more destinations that you would stream them to. Normal usage charges for the destination will occur. Learn
more about the different log categories and contents of those logs

Diagnostic setting name *   [diagnosticsettings01   ✓]

Logs   **Destination details**

Category groups ⓘ   ☑ Send to Log Analytics workspace
  ☐ allLogs   ☐ audit
                                                        Subscription
Categories   [Azure subscription 1   ∨]
  ☑ Connections
                                                        Log Analytics workspace
  ☑ Device Telemetry   [DefaultWorkspace-a8aa213d-fade-4421-b170-bbfa0c6edf41-EUS (eastus) ∨]

  ☐ C2D Commands   ☐ Archive to a storage account

  ☐ Device Identity Operations   ☐ Stream to an event hub

  ☐ File Upload Operations   ☐ Send to partner solution

  ☐ Routes

  ☐ D2CTwinOperations

  ☐ C2D Twin Operations

Figure 4.9: Selecting metrics to log

4. Send the logs to your Log Analytics Workspace.

5. Save the settings and close the **Diagnostic setting** pane.

6. In the **IoT Hub** menu, select **Metrics** from the **Monitoring** section.

7. In the upper-right corner above the graph, click on the time span and change it to **Last 4 hours**:

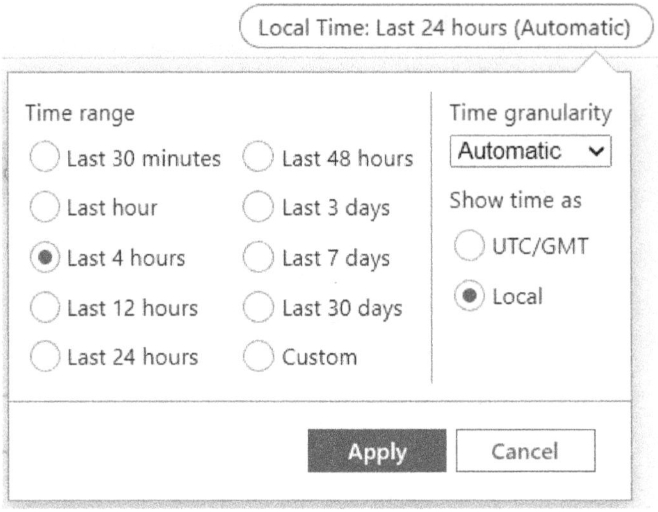

Figure 4.10: Selecting a time range

8. Make sure the scope for the metric is set to your IoT hub and the **Metric Namespace** option is set to **IoT Hub standard metrics**. For the **Metric** option, select **Telemetry messages sent** and an **Aggregation** value of **Sum**:

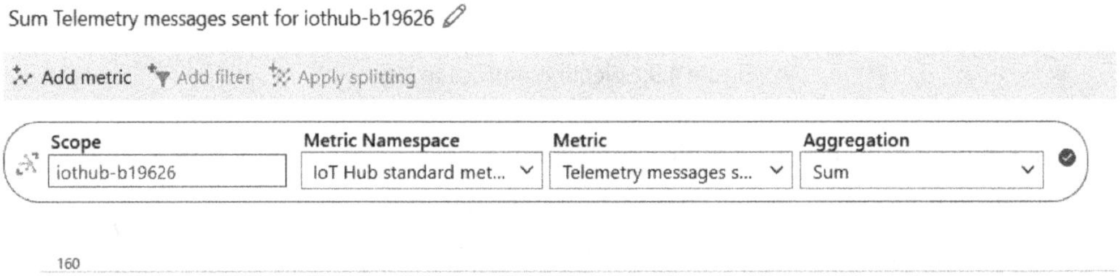

Sum Telemetry messages sent for iothub-b19626 ✎

Figure 4.11: Selecting an aggregation value

9. Repeat this procedure to add the **Total number of messages used** metric. Set the **Aggregation** value to **Avg**.

10. In the upper-right corner of the chart, select the **Save to dashboard** dropdown and select **Pin to dashboard**. Pin it to your default dashboard named `Dashboard`:

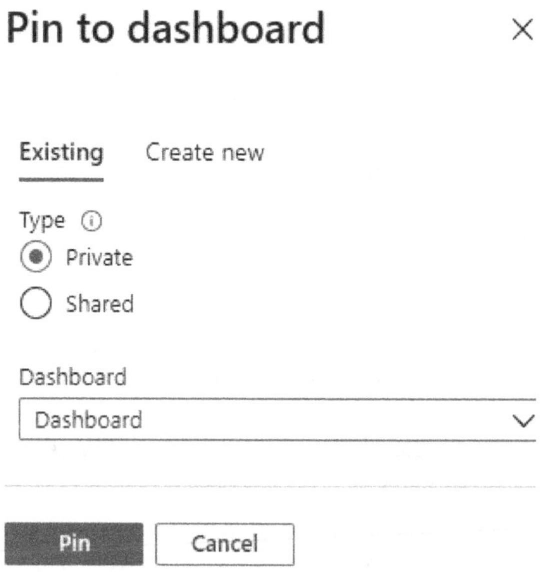

Pin to dashboard   ✕

Existing   Create new

Type ⓘ
● Private
○ Shared

Dashboard
┌──┐
│ Dashboard   ⌄ │
└──┘

[Pin]   [Cancel]

Figure 4.12: Pinning a chart to a dashboard

Now, we can set up metrics alerts:

1. Go to your **IoT Hub** blade in the Azure portal and select **Alerts** from the **Monitoring** section in the left-hand menu.

2. Select **Create alert rule**.

3. Configure the signal the alert will trigger on. In the **Select a signal** pane, select **Telemetry messages sent**.

4. In the **Configure signal logic** pane, set or confirm the following fields under **Alert logic**:

Parameter	Value
Threshold	`Static`
Aggregation type	`Total`
Operator	`Greater than`
Unit	`Count`
Threshold value	`200`
Check every	`1 minute`
Lookback period	`5 minutes`

Table 4.1: Parameter settings

5. On the **Action** tab, select **Create Action group**. Under the **Basics** tab, set the **Action group name** value and the **Display name** value to **Send Mail**.

6. Select the **Notification** tab. Select the **Notification type** dropdown and then select **Email/SMS message/Push/Voice**. In the **Email/SMS message/Push/Voice** pane, select **Email** and enter your email address.

7. On the **Actions** tab, you can select various actions to take. For this alert, we are only going to use notifications, so do not pick an action.

8. On the **Review and Create** tab, select **Create**.

9. Back in the **Create an alert rule** pane, select the **Details** tab. Give **Alert rule name** a value of **Alert me**.

10. Select the **Review and Create** tab. Select **Create**.

11. Set up another alert for **Total number of messages used**. Set it to alert you when the total number exceeds `200`. Check every minute with a lookback period of `1` minute.

12. If you go to the **Alerts** page, under **Monitoring** on the left-hand menu of the **IoT Hub** page you can view your alerts, as shown in the following screenshot. You may have to refresh the page a couple of times:

Home > iothub-b19626 | Alerts >

Alert rules   ...

+ Create   ≡≡ Columns   ↻ Refresh   ↓ Export to CSV   ⚬ Open query   🗑 Delete   ▷ Enable   ☐ Disable

☐ Search				
Subscription : **Azure subscription 1**			Signal type : **all**	

Name ↑↓	Condition	Severity ↑↓	Target scope	Target resource type
☐ Alert Me	d2c.telemetry.ingress.success >...	▌3 - Informational	iothub-b19626	IoT Hub
☐ Total Messages Used	dailyMessageQuotaUsed > 200	▌3 - Informational	iothub-b19626	IoT Hub

Figure 4.13: Viewing alerts

Run the simulated sensor, as follows:

1. Go to `https://azure-samples.github.io/raspberry-pi-web-simulator/#GetStarted`. You should see a Raspberry Pi Azure device simulator.

2. On *line 15* of the code editor, paste the connection string you saved earlier.

3. On *line 123*, you can set the message interval to `1000` to speed up the number of messages sent.

4. Let it run until the alerts are triggered. You should get an email when an alert is triggered. You can stop the simulator.

5. Go to the chart you pinned earlier; you should see metric data on the chart.

6. Go to the **Alerts** page for the IoT hub. You should see the alerts that have been fired.

7. Click on an alert name in the list. You should see more information on why the alert was fired.

8. Go to the **Logs** page for the IoT hub, select the **Recently connected devices** query, and run it. You should see your device ID listed:

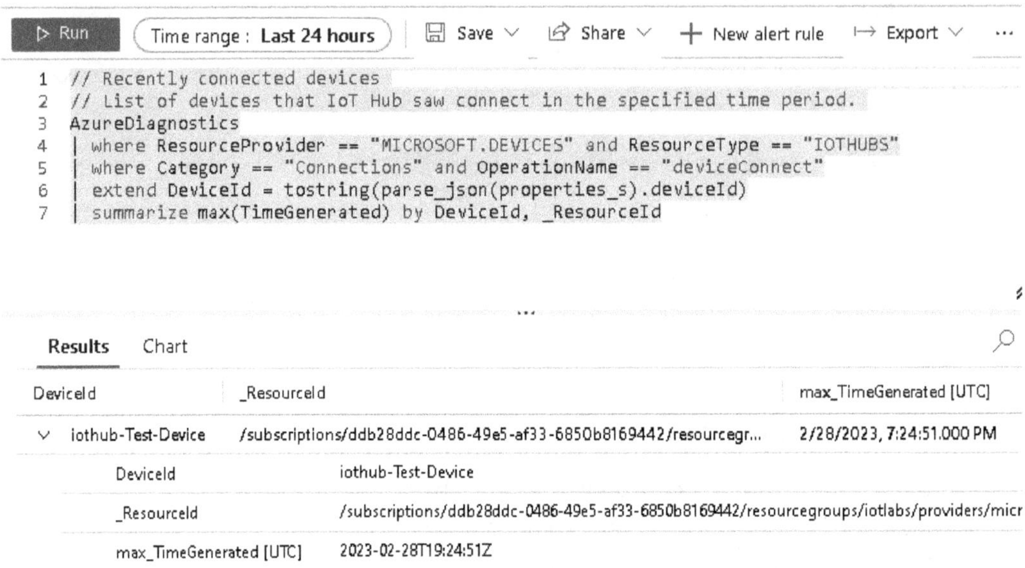

```
▷ Run     ( Time range :  Last 24 hours )    🖫 Save ⌄    🖻 Share ⌄    + New alert rule    ↦ Export ⌄    ⋯

1   // Recently connected devices
2   // List of devices that IoT Hub saw connect in the specified time period.
3   AzureDiagnostics
4   | where ResourceProvider == "MICROSOFT.DEVICES" and ResourceType == "IOTHUBS"
5   | where Category == "Connections" and OperationName == "deviceConnect"
6   | extend DeviceId = tostring(parse_json(properties_s).deviceId)
7   | summarize max(TimeGenerated) by DeviceId, _ResourceId
```

Results   Chart

DeviceId	_ResourceId	max_TimeGenerated [UTC]
⌄ iothub-Test-Device	/subscriptions/ddb28ddc-0486-49e5-af33-6850b8169442/resourcegr...	2/28/2023, 7:24:51.000 PM
DeviceId	iothub-Test-Device	
_ResourceId	/subscriptions/ddb28ddc-0486-49e5-af33-6850b8169442/resourcegroups/iotlabs/providers/micr	
max_TimeGenerated [UTC]	2023-02-28T19:24:51Z	

Figure 4.14: Querying the logs

9.   Clean up your resources.

Summary

In this chapter, the focus was on device management—specifically on how Azure IoT Hub streamlines the management of extensive device sets by automating tasks that are both complicated and repetitive. The significance of device twins as a fundamental tool for automating device management was also explored, along with the essential monitoring and logging necessary for establishing a secure IoT system.

In the next chapter, you will look deeper into security best practices, looking at Defender for IoT and how that can help to further secure your IoT solution.

5
Securing IoT Systems

As organizations and individuals embrace the potential of IoT, they are confronted with the daunting task of safeguarding their interconnected ecosystems from an ever-expanding array of threats. The consequences of IoT security breaches can be devastating, ranging from data theft and privacy violations to compromised infrastructure and, in some cases, physical harm.

This chapter explores the critical topic of *end-to-end security for IoT*. We delve into the strategies and solutions necessary to protect IoT ecosystems comprehensively. We begin by examining the fundamental principles of IoT security and the unique challenges it presents. From there, we pivot our focus to two key pillars of IoT security: Microsoft Defender for IoT and Microsoft Defender for Cloud.

Microsoft Defender for IoT is a robust, industry-leading solution designed to provide real-time threat protection and security management for IoT devices and networks. With its advanced capabilities, it enables organizations to proactively detect, respond to, and mitigate security threats across their IoT infrastructure.

Microsoft Defender for Cloud, on the other hand, extends this protection into the cloud environment, ensuring that data and applications are secure throughout their life cycle. It offers a holistic approach to cloud security, integrating seamlessly with Microsoft's broader ecosystem of security tools.

In particular, we will cover the following topics:

- End-to-end security for IoT
- Microsoft Defender for IoT
- Microsoft Defender for Cloud
- Lab – creating a security alert

End-to-end security for IoT

IoT security is critically important because IoT devices and systems are increasingly becoming targets for cybercriminals. These devices often contain sensitive information and control critical infrastructure, making them attractive targets for attackers.

Azure IoT provides a comprehensive set of security features to ensure end-to-end security for IoT devices and solutions. The following are the key pieces of Azure IoT end-to-end security:

- **Device identity**: Azure IoT provides device authentication and authorization using X.509 certificates, symmetric keys, and other authentication mechanisms to ensure that only authorized devices can access the IoT solution.

- **Data encryption**: Azure IoT uses industry-standard encryption mechanisms to encrypt data in transit and at rest. This includes **Transport Layer Security/Secure Sockets Layer (TLS/SSL)**.

- Encryption for communication between devices and the cloud, and **Advanced Encryption Standard (AES)** encryption for data at rest (see `https://www.usnews.com/360-reviews/privacy/what-is-advanced-encryption-standard`).

- **Secure device communication**: Azure IoT provides secure device communication using protocols such as **Message Queuing Telemetry Transport (MQTT)**, **Advanced Message Queuing Protocol (AMQP)**, and **Hypertext Transfer Protocol with encryption (HTTPS)**, which are designed to provide secure and reliable communication between devices and the cloud.

- **Secure cloud communication**: Azure IoT ensures secure communication between the cloud and devices by providing secure access control mechanisms, firewall protection, and other security features.

- **Device management**: Azure IoT provides device management capabilities, such as device provisioning and firmware updates, which are critical to maintaining the security of IoT devices.

- **Threat detection**: Azure IoT includes built-in threat detection capabilities that can identify and respond to potential security threats in real time.

- **Compliance**: Azure IoT is compliant with various security and privacy regulations, including the **International Organization for Standardization (ISO)** *27001*, the **Health Insurance Portability and Accountability Act (HIPAA)**, and the **General Data Protection Regulation (GDPR)**, ensuring that your IoT solution meets the highest security standards.

Since these security features are incorporated into your Azure IoT solution, you can ensure end-to-end security and protect your IoT devices and data from potential security threats.

A use case to exemplify how Azure IoT end-to-end security components work together

For example, imagine a large industrial manufacturing facility that produces complex machinery. To enhance efficiency and reduce downtime, the facility has implemented IoT devices and sensors throughout its operations. These devices collect data on machine performance, environmental conditions, and production metrics, which is then analyzed in real time to optimize production processes.

One day, the industrial facility's security team receives an alert from Azure IoT security center indicating an anomaly in the behavior of a critical manufacturing machine. The machine's IoT sensor data shows a sudden drop in temperature readings, which could lead to equipment damage and production delays.

The security team takes the following actions:

1. **Alert investigation**: The security team accesses Azure IoT Security Center to investigate the alert further. They analyze historical data and device logs to understand what caused the temperature drop.

2. **Remote troubleshooting**: With Azure IoT Hub, the team remotely connects to the troubled IoT device and assesses its condition. They discover a malfunctioning sensor and initiate a remote firmware update to fix the issue.

3. **RBAC verification**: During the investigation, the security team ensures that only authorized personnel have access to the device and its control systems, preventing unauthorized tampering.

4. **Secure communication**: Data sent between the device and Azure IoT Hub remains encrypted, safeguarding sensitive information from eavesdropping.

5. **Incident resolution**: After the firmware update, the machine's temperature readings return to normal, averting potential damage and production delays.

This use case illustrates how Azure IoT end-to-end security components work together to detect and respond to a potential security issue, ensuring the integrity and reliability of industrial processes in a secure and efficient manner.

In summary, IoT security is crucial for protecting sensitive data, ensuring device safety, preventing device hijacking, maintaining customer trust, and complying with regulations. By implementing strong security measures, IoT device manufacturers and solution providers can help mitigate risks associated with IoT devices and ensure a secure and reliable IoT ecosystem. Let's look closer at some of the tools used for security in an Azure IoT system. We will start with Microsoft Defender for IoT.

Microsoft Defender for IoT

Microsoft Defender for IoT is a cloud-based security solution designed to help organizations discover, assess, and monitor the security posture of their IoT devices and systems. It provides threat protection for IoT devices and networks, giving organizations the visibility and control they need to secure their IoT environments.

Microsoft Defender for IoT works by integrating with existing IoT devices and networks, collecting data from devices, and analyzing it for potential security threats. It uses **machine learning** (**ML**) and advanced analytics to identify anomalies and patterns that could indicate a security breach.

Some key features of Microsoft Defender for IoT include:

- **Continuous monitoring**: Microsoft Defender for IoT provides real-time monitoring of IoT devices and networks, allowing organizations to detect and respond to security threats quickly

- **Threat detection**: The solution uses ML algorithms and behavioral analysis to identify potential security threats, such as malware infections, network intrusions, and unauthorized access

- **Vulnerability management**: Microsoft Defender for IoT assesses the security posture of IoT devices and provides recommendations for mitigating vulnerabilities

- **Incident response (IR)**: The solution provides detailed incident reports and actionable recommendations for responding to security threats

- **Integration with other Azure services**: Microsoft Defender for IoT integrates with other Azure services, such as Azure Sentinel and Azure Security Center, to provide a comprehensive security solution for organizations

Microsoft Defender for IoT is a powerful tool for securing IoT devices and networks, helping organizations identify and mitigate security threats before they can cause damage. By using this solution, organizations can maintain the security and reliability of their IoT environments and ensure the privacy and safety of their customers.

Setting up Microsoft Defender for IoT

By default, Microsoft Defender for IoT is enabled on all new instances of an Azure IoT hub, as seen here:

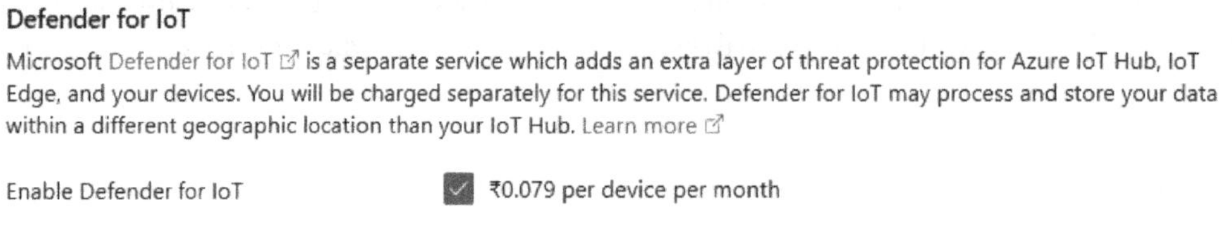

Defender for IoT

Microsoft Defender for IoT ↗ is a separate service which adds an extra layer of threat protection for Azure IoT Hub, IoT Edge, and your devices. You will be charged separately for this service. Defender for IoT may process and store your data within a different geographic location than your IoT Hub. Learn more ↗

Enable Defender for IoT   ☑ ₹0.079 per device per month

Figure 5.1 – Enabling Defender for IoT

Once the IoT hub is created, you should see a **Defender for IoT** section on the left side menu:

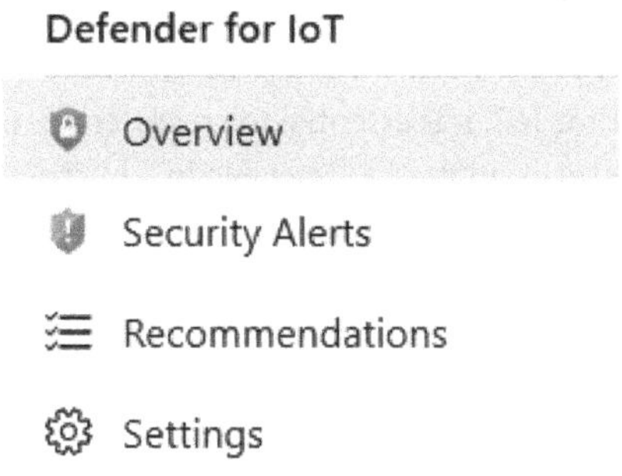

Defender for IoT

🛡 Overview

🛡 Security Alerts

☰ Recommendations

⚙ Settings

Figure 5.2 – Defender for IoT menu selections

As you can see in the preceding screenshot, IoT Hub provides agentless monitoring of your devices, along with the ability to set up alerts and recommendations for improving your security posture. It also integrates and interoperates with **Microsoft 365 Defender**, **Microsoft Sentinel**, and external **security operations center** (**SOC**) tools. Take some time to investigate the various pages under the **Defender for IoT** heading in the menu.

You can add security agents to your devices to get further insight into any threats on your devices. We will look at this next.

Using security agents

A **Defender for IoT micro agent** is a security software designed to protect IoT devices from cyber-attacks. Specifically, it works by continuously monitoring the device's activities and behavior for any signs of malicious activity or anomalies that could indicate a security breach.

The Defender for IoT micro agent typically employs a combination of ML and behavioral analysis techniques to detect and prevent various types of cyber-attacks, such as malware, ransomware, and botnet attacks. It may also include features such as intrusion detection and prevention, network segmentation, and access control to further strengthen the security of IoT devices.

Overall, the primary function of a Defender for IoT micro agent is to ensure the security and integrity of IoT devices and the data they process, by proactively identifying and mitigating security threats before they can cause any harm.

In Azure IoT solutions, **device twins** are essential for both device management and process automation. The Defender for IoT solution allows integration with your existing IoT device management platform, which enables you to manage device security and use existing device control capabilities. The **IoT Hub twin mechanism** is used to integrate your Defender for IoT solution with your device management platform.

The module twin mechanism is used by Defender for IoT, which maintains a specific twin called `DefenderIotMicroAgent` for each device. This twin is a digital replica of the device and contains all the necessary data about its behavior and security status.

To use all the features of Defender for IoT, it is essential to create, configure, and utilize Defender-IoT-micro-agent twins. The following is an example of a Defender-IoT-micro-agent twin. In this example, the Defender-IoT-micro-agent twin contains information about the security status and network interfaces of an IoT device called `myIoTDevice`. The `desired` section specifies the initial configuration of the Defender-IoT-micro-agent twin, while the `reported` section contains the current status of the twin. Here's the sample `DefenderIotMicroAgent` device twin:

```json
{
    "deviceId": "myIoTDevice",
    "etag": "AAAAAAAAAAU=",
    "properties": {
      "desired": {
        "DefenderIotMicroAgent": {
          "deviceId": "myIoTDevice",
          "securityStatus": {
            "antivirusStatus": "UpToDate",
            "firewallStatus": "Enabled"
          },
  . . .


  . . .      "reported": {
        "DefenderIotMicroAgent": {
          "lastUpdated": "2023-04-05T13:45:30.000Z",
          "securityStatus": {
            "antivirusStatus": "UpToDate",
            "firewallStatus": "Enabled",
            "intrusionDetectionStatus": "Enabled"
          }
        }
      }
    }
}
```

The security status of the device is listed as `UpToDate` for antivirus and `Enabled` for firewall, indicating that the device has the latest antivirus definitions and the firewall is turned on. Additionally, intrusion detection is listed as `Enabled` in the `reported` section, which indicates that the Defender for IoT solution has enabled this feature on the device.

The network interfaces of the device are also listed, including the IPv4 address, MAC address, **network security groups** (**NSGs**), and virtual network configuration. This information can be used by the Defender for IoT solution to monitor network traffic and detect any suspicious activity.

Another security solution is Microsoft Defender for Cloud. Let's see how this fits in next.

Microsoft Defender for Cloud

Microsoft Defender for Cloud and Microsoft Defender for IoT are both security solutions that provide protection for different types of environments. Microsoft Defender for Cloud is a **cloud access security broker** (**CASB**) solution that provides **advanced threat protection** (**ATP**), data protection, and governance for cloud applications and services. On the other hand, Microsoft Defender for IoT is an endpoint protection platform that provides security for IoT devices.

While these two solutions address different areas of security, they can complement each other in a few ways:

- **Improved visibility**: Microsoft Defender for Cloud can provide visibility into cloud applications and services used by IoT devices, while Microsoft Defender for IoT can provide visibility into the security status of the IoT devices themselves. Together, these solutions can offer a more comprehensive view of the security posture of an organization's entire digital estate.

- **Threat detection and response**: Microsoft Defender for Cloud can detect and respond to threats within cloud environments, while Microsoft Defender for IoT can detect and respond to threats on IoT devices. By integrating these solutions, organizations can quickly detect and respond to attacks that span both cloud and IoT environments.

- **Integrated security management**: Microsoft Defender for Cloud and Microsoft Defender for IoT can be managed through a single console, allowing organizations to easily manage their security policies and settings for both cloud and IoT environments.

- **Automated remediation**: With integration, these solutions can automatically remediate threats detected in one environment (for example, cloud) by taking action in the other environment (for example, IoT).

By working together, Microsoft Defender for Cloud and Microsoft Defender for IoT can provide a more comprehensive and integrated security solution that covers cloud and IoT environments, reducing the risk of attacks and protecting an organization's critical assets.

> **Threat modeling is a crucial part of IoT security**
>
> Microsoft's Threat Modeling Tool serves as a fundamental component of the Microsoft **Security Development Lifecycle** (**SDL**), allowing software architects to proactively identify and address potential security issues in their early stages when they are more manageable and cost-effective to resolve. Consequently, this approach significantly reduces overall development costs. Moreover, they have designed the tool with the intention of making threat modeling more accessible to all developers, regardless of their security expertise, by providing clear guidance on how to create and analyze threat models. This Microsoft tool employs the STRIDE model, which categorizes various threat types and simplifies discussions about security:
>
> - **Spoofing**: Involves unauthorized access and use of another user's authentication information, such as usernames and passwords.
>
> - **Tampering**: Refers to the malicious alteration of data, including unauthorized changes to persistent data stored in a database and data modification during transmission over a network, such as the internet.
>
> - **Repudiation**: Concerns users who deny their involvement in an action without any means for others to prove otherwise, such as a user performing an unlawful operation in a system lacking traceability. Non-repudiation relates to a system's ability to counter repudiation threats, such as requiring a user to sign for a received item as proof of delivery.
>
> - **Information disclosure**: Involves the exposure of information to unauthorized individuals, such as users reading files they lack access to or intruders intercepting data in transit between computers.
>
> - **Denial of service**: **Denial-of-service** (**DoS**) attacks disrupt services for legitimate users, rendering a web server temporarily unavailable or unusable, necessitating protection to enhance system availability and reliability.
>
> - **Elevation of privilege**: This occurs when an unprivileged user gains elevated access, potentially compromising or jeopardizing the entire system. Elevation of privilege threats encompass scenarios where attackers effectively breach all system defenses and become part of the trusted system itself – a highly precarious situation.

Now that we have discussed the theory behind IoT security, it is time to roll up your sleeves and implement Microsoft Defender for IoT in the following lab.

Lab – creating a security alert

This lab shows how to enable Defender for IoT and configure data collection. You will then install and use a micro-agent to collect security data. At the end of this lab, you should be able to:

- Enable Defender for IoT

- Collect security data using Defender for IoT

Let's get started with the lab:

1. Log in to your Azure portal.

2. Create a resource group and an IoT hub inside the resource group. On the **Add-ons** tab, make sure **Enable Defender for IoT** is checked:

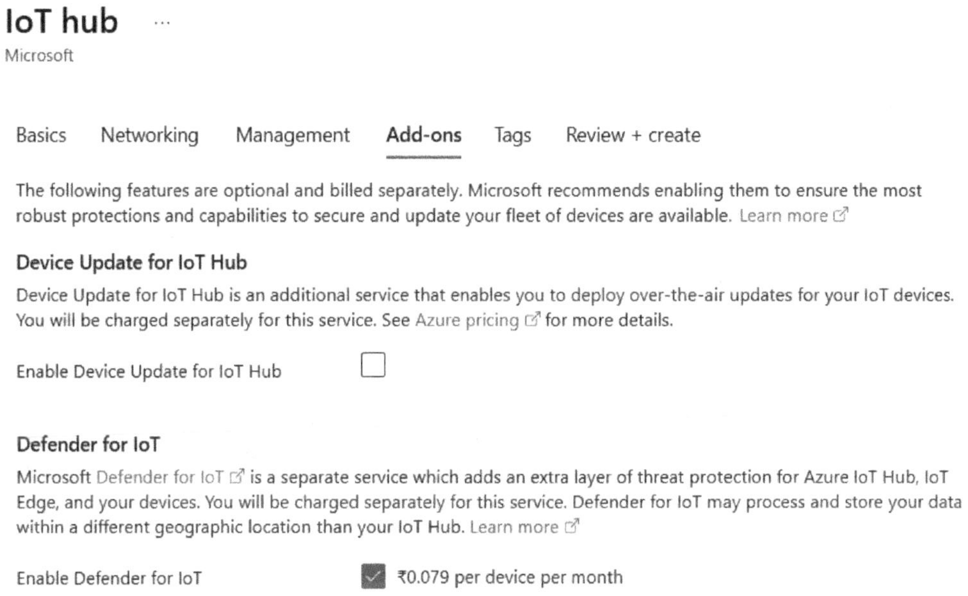

IoT hub   ...
Microsoft

Basics   Networking   Management   **Add-ons**   Tags   Review + create

The following features are optional and billed separately. Microsoft recommends enabling them to ensure the most robust protections and capabilities to secure and update your fleet of devices are available. Learn more ☐

Device Update for IoT Hub

Device Update for IoT Hub is an additional service that enables you to deploy over-the-air updates for your IoT devices. You will be charged separately for this service. See Azure pricing ☐ for more details.

Enable Device Update for IoT Hub   ☐

Defender for IoT

Microsoft Defender for IoT ☐ is a separate service which adds an extra layer of threat protection for Azure IoT Hub, IoT Edge, and your devices. You will be charged separately for this service. Defender for IoT may process and store your data within a different geographic location than your IoT Hub. Learn more ☐

Enable Defender for IoT   ☑   ₹0.079 per device per month

Figure 5.3 – Enabling Defender for IoT

3. Once created, go to the **Defender for IoT** section on the left menu and select **Settings**.

4. Under **Data Collection**, select **Workspace configuration** to configure a workspace for your security logs to be sent.

5. Once the workspace is created, go back to the IoT hub and the **Data Collection** option under the **Defender for IoT** section of the side menu. Update **Workspace configuration** to point to the new workspace, and make sure the **Access to raw security data, In-depth security recommendations and custom alerts**, and **IP data collection** checkboxes are all checked:

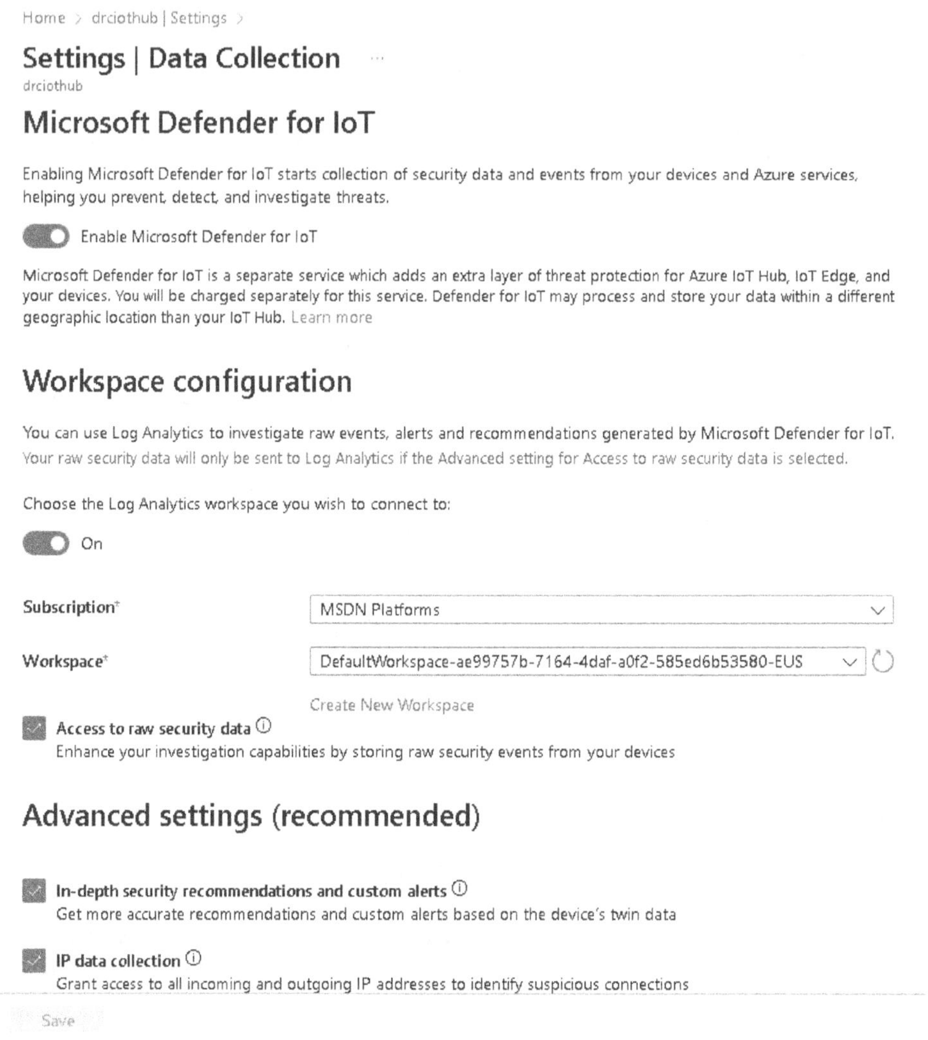

Home > drciothub | Settings >

Settings | Data Collection
drciothub

Microsoft Defender for IoT

Enabling Microsoft Defender for IoT starts collection of security data and events from your devices and Azure services, helping you prevent, detect, and investigate threats.

🔘 Enable Microsoft Defender for IoT

Microsoft Defender for IoT is a separate service which adds an extra layer of threat protection for Azure IoT Hub, IoT Edge, and your devices. You will be charged separately for this service. Defender for IoT may process and store your data within a different geographic location than your IoT Hub. Learn more

Workspace configuration

You can use Log Analytics to investigate raw events, alerts and recommendations generated by Microsoft Defender for IoT. Your raw security data will only be sent to Log Analytics if the Advanced setting for Access to raw security data is selected.

Choose the Log Analytics workspace you wish to connect to:

🔘 On

Subscription*	MSDN Platforms ⌄
Workspace*	DefaultWorkspace-ae99757b-7164-4daf-a0f2-585ed6b53580-EUS ⌄ ↻

Create New Workspace

☑ **Access to raw security data** ⓘ
Enhance your investigation capabilities by storing raw security events from your devices

Advanced settings (recommended)

☑ **In-depth security recommendations and custom alerts** ⓘ
Get more accurate recommendations and custom alerts based on the device's twin data

☑ **IP data collection** ⓘ
Grant access to all incoming and outgoing IP addresses to identify suspicious connections

Save

Figure 5.4 – Configuring Defender for IoT settings

6. Save your changes, go back to your IoT hub, and register a new device called `device1`. Use the **Symmetric key** option for the **Authentication type** field.

7. Once the device is registered, copy the **Primary connection string** value for later use.

8. To simulate a device, go to `https://azure-samples.github.io/raspberry-pi-web-simulator`, and on line 15 of the code, enter the connection string from *step 7*.

9. Run the simulator and verify that you are getting messages in your IoT hub.

10. In the Azure portal, navigate to your IoT hub. On the left side menu, under **Defender for IoT**, click **Settings**:

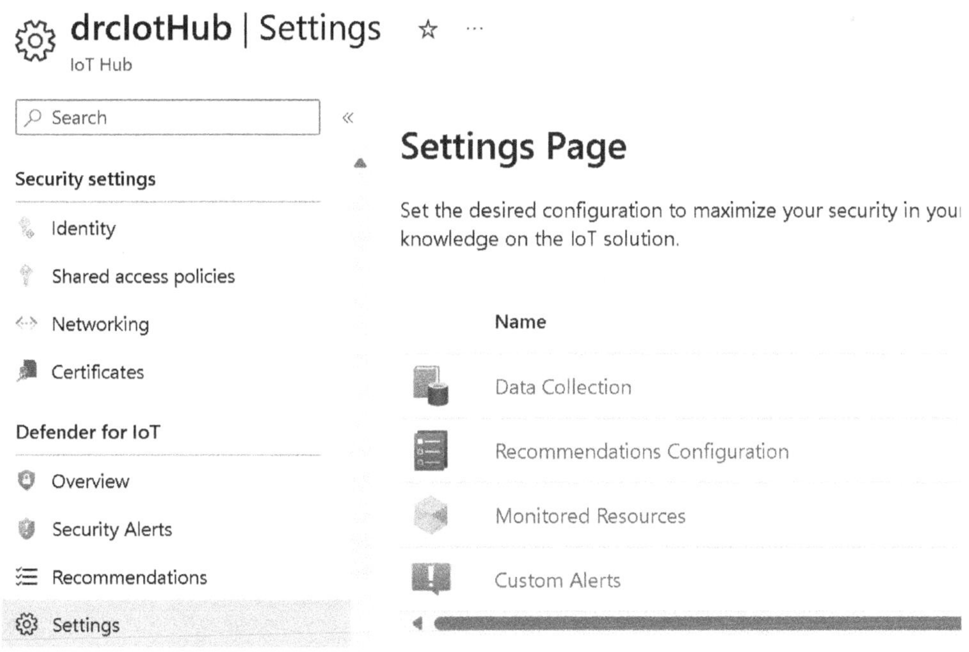

drclotHub | Settings ☆ ⋯
IoT Hub

🔍 Search   «

Security settings
─────────────
🔖 Identity

🔑 Shared access policies

↔ Networking

🛡 Certificates

Defender for IoT
─────────────
🛡 Overview

🛡 Security Alerts

☰ Recommendations

⚙ Settings

Settings Page

Set the desired configuration to maximize your security in your
knowledge on the IoT solution.

Name

▢ Data Collection

▢ Recommendations Configuration

⬡ Monitored Resources

▢ Custom Alerts

◀ ▬▬▬▬▬▬▬▬▬▬▬▬▬▬▬▬

Figure 5.5 – Setting up the monitored resources

11. Under **Settings**, click **Monitored Resources**. At the top of the pane, click **Edit**.

12. In the **Solution Management** pane, select your subscription and resource group:

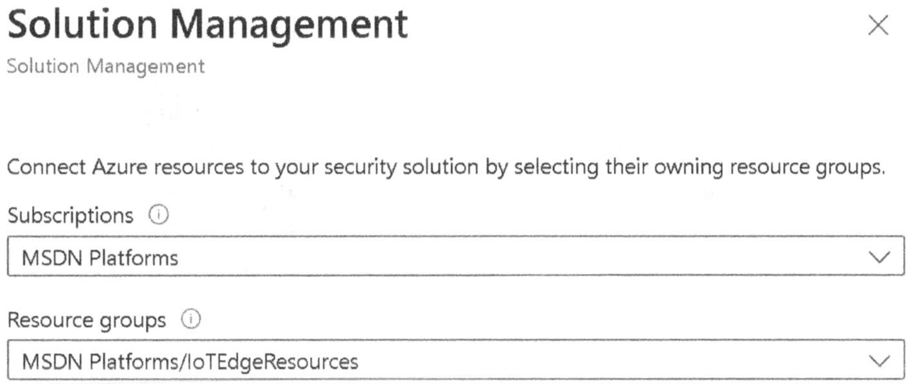

Solution Management   ✕
Solution Management

Connect Azure resources to your security solution by selecting their owning resource groups.

Subscriptions ⓘ

MSDN Platforms	∨

Resource groups ⓘ

MSDN Platforms/IoTEdgeResources	∨

Figure 5.6 – Connecting Azure resources

13. Back on the IoT hub, under the left side menu, select **Overview**. Review the **Threat prevention** and **Threat detection** information presented in the pane:

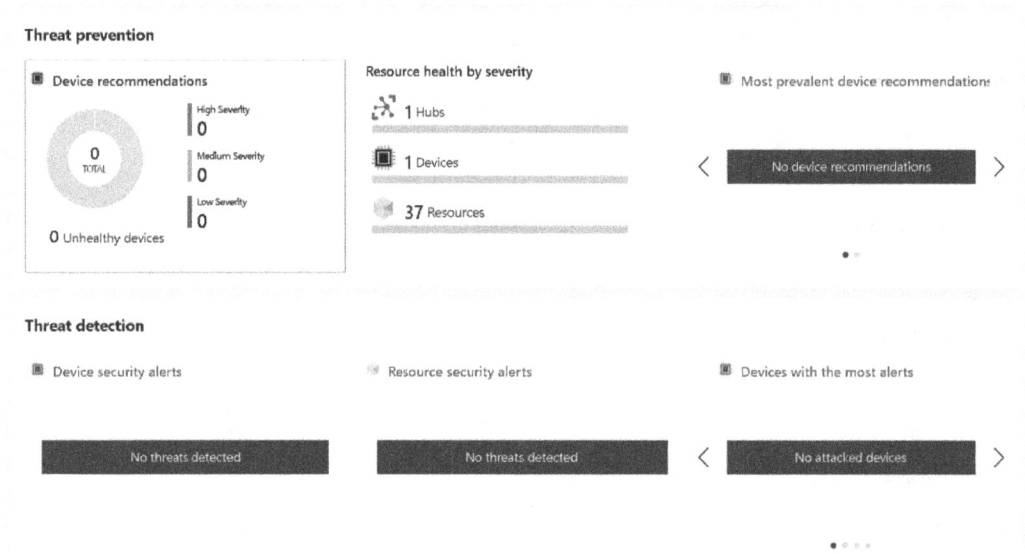

Threat prevention

Threat detection

Figure 5.7 – Observing threat detection

14. To create a custom alert, go to your IoT hub and select **Settings** under the **Defender for IoT** section in the menu. On the **Settings** page, select **Custom Alerts**.

15. On the **Custom Alerts** page, you should see a default security group listed.

16. Click on the default security group. At the top of the default security group pane, select **Create custom alert rule**.

17. In the **Create custom alert rule** pane, there should be a **Custom Alert** dropdown. Under the dropdown, select **Number of cloud to device messages (MQTT protocol) is not in allowed range**.

18. For the **Minimal Threshold** field, enter 1, and for **Maximal Threshold**, enter 10. For the **Time Window Size** field, select 5 minutes (00:05:00). Click the **OK** button and make sure to save the alert rule:

Create custom alert rule

Create custom alert rule

Custom Alert Type

Device Security Group

default

Custom Alert *

Number of cloud to device messages (MQTT protocol) is not in allowed range

Description

Get an alert when the number of cloud to device messages (MQTT protocol) in the time window is not in the allowed range

Required Properties

Minimal Threshold * ⓘ

1

Maximal Threshold * ⓘ

10

Time Window Size ⓘ

00:05:00

Figure 5.8 – Creating a custom alert rule

19. To add the device you registered earlier to the default security group, alter the device twin by adding the following between the version and property fields: `"tags" : {"SecurityGroup": "default"},`

20. If you don't have the simulated device page still open, go back to `https://azure-samples. github.io/raspberry-pi-web-simulator` and on line 15 of the code, enter the connection string from *step 7*.

21. On the simulated device page, select **Run** to start sending signals.

22. Select the **Overview** page under the **Defender for IoT** group on the IoT hub left side menu. You should start seeing device alerts.

> **Note**
>
> While writing this chapter, Microsoft was changing the way it was implementing IoT security. You should verify any changes and use the latest approved methods and services. You can keep up with the changes and updates at `https://techcommunity.microsoft.com/t5/internet-of-things-blog/bg-p/IoTBlog`.

Now that you have completed this chapter and lab, let's look at what we have learned.

Summary

This chapter emphasized the importance of securing IoT systems due to their inherent complexity. It discussed the need to protect various aspects of the system. The chapter provided an introduction to ensuring system security and the various tools that can simplify the process. Key topics covered include end-to-end security for IoT, Microsoft Defender for IoT, utilizing security agents, Microsoft Defender for Cloud, and a lab session on Microsoft Defender for IoT. Although this chapter just covered the basics, security is an important topic, and you should work with your cloud security experts to make sure your IoT security plan is up to date and robust.

Following the acquisition of streaming data from your devices, the subsequent phase involves data processing to extract valuable insights. Over the upcoming chapters, we will delve into various aspects such as message routing, long-term trend analysis through message storage, real-time streaming analytics with **Azure Stream Analytics** (**ASA**), and time-series insights utilizing Azure Data Explorer. Additionally, we will explore the potential of decentralizing certain analytics processing tasks onto the devices themselves.

At the outset of signal processing, an initial task involves directing these signals toward different destination points. In the next chapter, you will uncover the fundamental principles underpinning effective message routing. You will also become acquainted with the flexible customization choices at your disposal, allowing you to tailor communication routes according to your precise requirements. Moreover, we will explore the impactful concept of message routing queries, which provide you with deeper insights and enhanced control over the flow of your data. Additionally, we will immerse ourselves in the transformative practice of message enrichment. Through this, you will gain proficiency in augmenting your data payloads with valuable contextual information, thereby opening avenues for more comprehensive and well-informed decision-making.

Part 2: Processing the Data

The second part of this book deals with processing your data after it has been collected; this includes routing messages to various endpoints, depending on how they needs to be processed. We will look at real-time processing of streamed data with Azure Stream Analytics. We will also investigate using Azure Data Explorer to process the data. This part concludes with a discussion of IoT edge devices and moving some data processing to devices on the edge.

This part has the following chapters:

- *Chapter 6, Creating Message Routing*
- *Chapter 7, Exploring Azure Stream Analytics*
- *Chapter 8, Investigating IoT Data with Azure Data Explorer*
- *Chapter 9, Exploring IoT Edge Computing*

6

Creating Message Routing

In the previous chapters, you have seen how to set up, deploy, and secure your IoT devices. You have also sent signals from devices to an IoT hub. Now that you can set up IoT systems, it is time to learn how to process these signals. One of the first steps in processing signals is to route them to various endpoints. As we embark on this exploration, you will discover the key principles behind efficient message routing, the customization options available to tailor communication pathways to your specific needs, and how to leverage message routing queries to gain deeper insights and control over your data flows. Additionally, we will delve into the transformative practice of message enrichment, where you will learn how to enhance your data payloads with valuable contextual information, opening the door to richer and more informed decision-making.

In particular, we will look at the following topics:

- Message routing basics
- Built-in endpoints
- Custom endpoints
- Message routing queries
- Message enrichment
- Lab – using message enrichment and custom endpoints

Exploring the basics and overall process of message routing

In an IoT hub, **message routing** involves the process of directing and delivering messages between IoT devices and cloud-based services or other endpoints. The IoT hub acts as a central hub or intermediary that facilitates communication and data exchange in an IoT ecosystem. Here's an overview of the message routing process:

1. **Device connectivity**: IoT devices, such as sensors or actuators, establish a connection with the IoT hub. They can use various communication protocols, including **Message Queuing Telemetry Transport (MQTT)**, **Advanced Message Queuing Protocol (AMQP)**, or **Hypertext Transfer Protocol (HTTP)**, to connect to the IoT hub over the internet or other supported networks.

2. **Message ingestion**: The IoT hub receives messages from the connected devices. These messages can contain sensor data, device telemetry, commands, or other types of information.

3. **Routing rules**: Within the IoT hub, you can define routing rules that specify how messages should be routed based on certain criteria. Routing rules are configured using a rule engine provided by the IoT Hub service. The criteria can include message properties, device metadata, message content, or other attributes.

4. **Message evaluation**: When a message arrives at the IoT hub, the routing rules are evaluated to determine the appropriate routing path for the message. The rules can be based on conditions, filters, or transformations that you define. For example, you can route messages based on the device ID, message type, or payload content.

5. **Endpoint selection**: Based on the evaluation of the routing rules, the IoT hub selects the appropriate endpoint or destination for the message. Endpoints can include other cloud services, storage systems, databases, or even other IoT devices. The IoT Hub service has built-in integration capabilities to connect with popular cloud platforms and services.

6. **Message delivery**: Once the endpoint is determined, the IoT hub forwards the message to the selected destination. The message can be delivered to cloud-based services for further processing, stored in a database for later retrieval, or sent to other IoT devices for coordination and interaction.

7. **Acknowledgment and feedback**: After the message is successfully delivered to the destination, the IoT hub can provide acknowledgments or feedback to the device. This feedback can include status updates, response messages, or instructions for the device to take appropriate actions.

By leveraging routing rules and flexible endpoint configurations, an IoT hub enables intelligent and dynamic message routing within an IoT ecosystem. It allows you to efficiently manage and process large volumes of data generated by IoT devices and route them to the appropriate destinations for analysis, storage, visualization, or triggering further actions.

If you go to the **Hub settings** section of the left side menu of your IoT hub, you will see a built-in endpoint, which is an Event Hub-compatible endpoint. This is used for system and device messaging. If you select this, you will see the following:

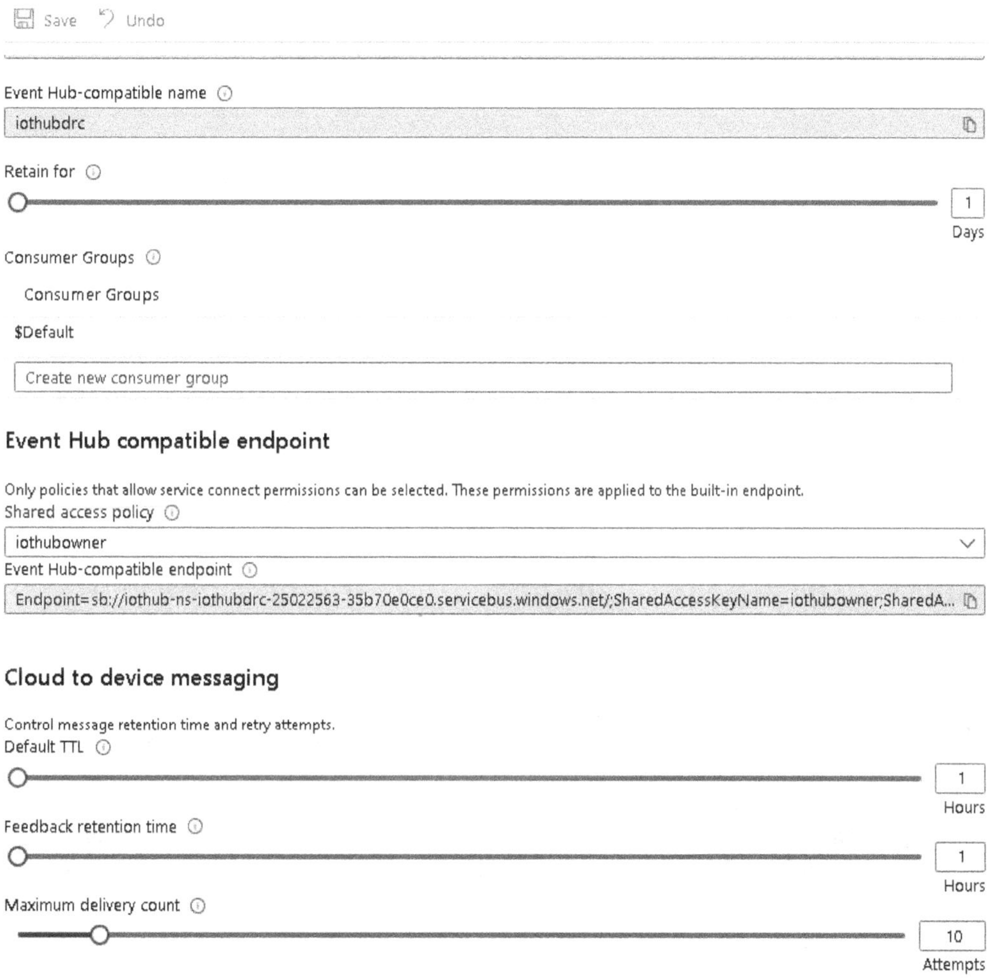

Figure 6.1 – Built-in endpoint

As you can see, there is a **Cloud to device messaging** section where you can set things such as feedback retention, retry attempts, and delivery count.

Under the same **Hub settings** section on the left side menu, you can select **Message routing**, where you can add routes.

When you add a route, the first step is to add an endpoint. You can use the **built-in endpoint** or a **custom endpoint**. If you select the dropdown for the various endpoints, you can see some of the more common ones used:

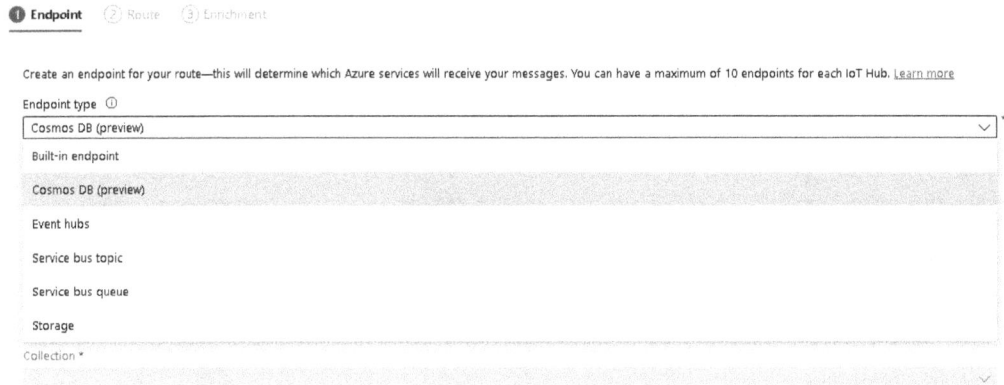

Figure 6.2 – Selecting an endpoint

Let's have a closer look at built-in endpoints.

Common use cases for using the built-in endpoint in an IoT hub

The built-in endpoint in an IoT hub provides a convenient way to route messages from IoT devices to different backend systems or applications. Here are some common use cases for using the built-in endpoint in an IoT hub:

- **Telemetry data storage**: IoT devices often generate large volumes of telemetry data. The built-in endpoint can be used to route telemetry messages to a storage system such as Azure Blob Storage or Azure Data Lake Storage for long-term storage and analysis.

- **Real-time analytics**: The built-in endpoint can forward telemetry data to real-time analytics systems such as **Azure Stream Analytics** (**ASA**) or Azure Databricks. This enables you to perform real-time processing, aggregations, and analytics on incoming data.

- **Command and control**: IoT Hub allows bidirectional communication with devices. The built-in endpoint can be used to receive command messages from backend systems or applications and route them to the appropriate devices for control and configuration purposes.

- **Rule-based message routing**: You can configure IoT Hub to route messages based on predefined rules. The built-in endpoint can be used to define rules that determine how messages are filtered and forwarded to different endpoints or systems based on message content, device metadata, or other criteria.

- **Integration with Azure Functions or Logic Apps**: The built-in endpoint can forward messages to Azure Functions or Azure Logic Apps, allowing you to trigger serverless functions or workflows in response to specific events or conditions. This enables you to implement custom business logic and automate actions based on incoming data.

- **Event processing and notifications**: The built-in endpoint can route messages to Azure Event Grid, which provides event-based routing and processing capabilities. You can use Event Grid to trigger event-driven workflows, send notifications, or integrate with other Azure services and third-party systems.

- **Data transformation**: IoT Hub supports message enrichment and transformation through various features such as message routing, custom endpoints, and Azure Functions integration. The built-in endpoint can be used in conjunction with these features to transform, enrich, or normalize incoming data before forwarding it to other systems.

These are just a few examples of the use cases for using the built-in endpoint in an IoT hub. The flexibility and extensibility of IoT Hub allow you to adapt the solution to your specific needs and integrate with a wide range of Azure services and external systems.

In the next section, we will take a closer look at custom endpoints.

Common use cases for using custom endpoints in an IoT hub

Using a custom endpoint in an IoT hub offers additional flexibility and control over where messages from IoT devices are sent. Here are some use cases for using a custom endpoint in an IoT hub:

- **Custom data processing**: By setting up a custom endpoint, you can route messages from IoT devices directly to your own backend systems for custom data processing. This enables you to implement specialized data processing logic, perform complex analytics, or integrate with existing enterprise systems.

- **Legacy system integration**: If you have legacy systems or on-premises infrastructure that needs to consume IoT device data, a custom endpoint can be configured to forward messages to these systems. This allows you to seamlessly integrate IoT data into your existing workflows and applications.

- **Third-party service integration**: IoT Hub's custom endpoint can be used to integrate with third-party services or platforms that require specific data formats or protocols. For example, you can forward messages to external analytics tools, **machine learning** (**ML**) services, or industry-specific platforms to leverage their capabilities for advanced data analysis and decision-making.

- **Edge computing integration**: IoT Hub can be used in conjunction with Azure IoT Edge to extend cloud capabilities to the edge of the network. With a custom endpoint, you can route messages from IoT devices to edge devices running IoT Edge modules for local data processing, real-time analytics, and decision-making closer to the source of data.

- **Data backup and redundancy**: A custom endpoint allows you to create redundant data storage or backup mechanisms for IoT device messages. You can configure the endpoint to send copies of the messages to alternative storage systems or replicate the data across multiple locations to ensure data integrity and availability.

- **Regulatory compliance**: In certain industries, there might be regulatory requirements or data sovereignty concerns that dictate where and how IoT device data is processed and stored. A custom endpoint provides the flexibility to route messages to specific regions or comply with data governance regulations.

- **Custom authentication and security**: By using a custom endpoint, you can implement your own authentication and security mechanisms tailored to your specific requirements. This allows you to enforce additional security measures, such as custom token-based authentication or encryption protocols, for securing communication between IoT devices and your backend systems.

These are just a few examples of use cases for using a custom endpoint in an IoT hub. The custom endpoint feature enables you to integrate IoT data with your infrastructure, applications, and services in a way that aligns with your unique needs and architecture.

The following screenshot shows the selection of a blob storage as the endpoint:

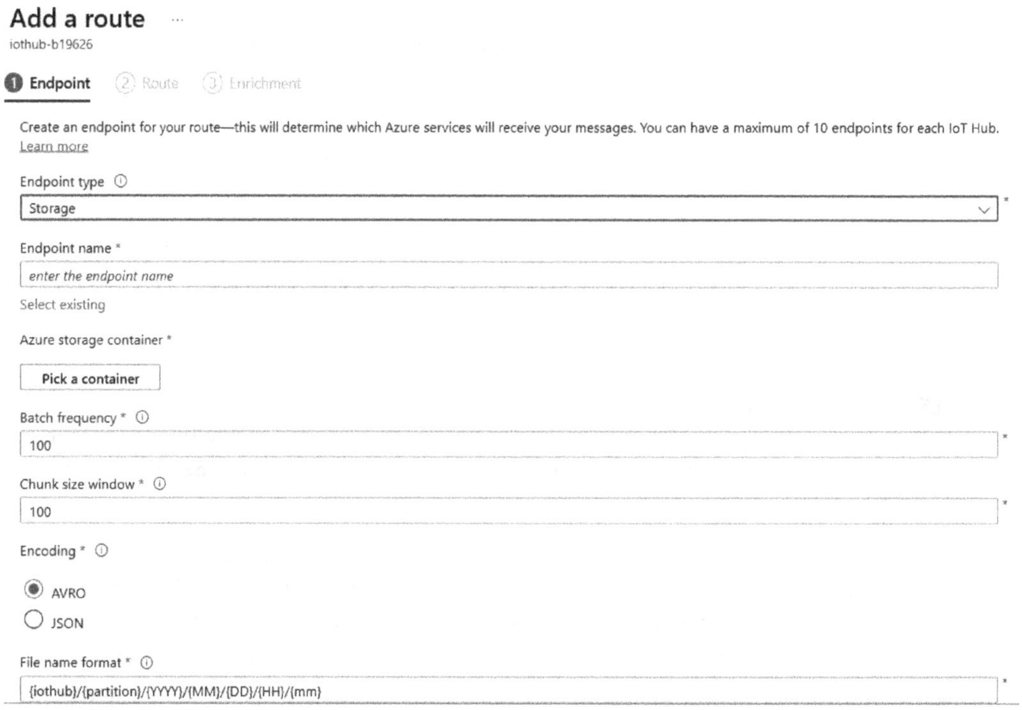

Figure 6.3 – Setting up a blob storage endpoint

After setting up the endpoint, the next step is to set up a route to send messages to the endpoint. We will look at this next.

Message routing queries

In the context of an IoT hub, **routing** refers to the process of directing and forwarding incoming messages from IoT devices to different endpoints or destinations within the IoT ecosystem. IoT Hub is a cloud-based service that acts as a central hub for managing and communicating with many connected devices.

Routing in an IoT hub enables you to define rules and conditions based on the message content or device metadata and then automatically route the messages to appropriate endpoints. These endpoints can include other Azure services, custom applications, storage systems, databases, or even other IoT hubs.

Here's a high-level overview of how routing works in an IoT hub:

1. **The device sends a message**: An IoT device connected to the IoT hub sends a telemetry message, event, or state update to the hub.

2. **Routing rules**: The IoT hub evaluates the routing rules defined by the user. These rules specify conditions and filters based on message properties such as message content, device metadata, or message properties added by device twins.

3. **Matching rules**: The IoT hub matches the incoming message against the defined routing rules to determine which rule(s) it satisfies.

4. **Destination determination**: Based on the matched rules, the IoT hub determines the destination(s) for the message. The destinations can be other IoT hubs, Azure services such as Event Hubs, Service Bus, or Event Grid, storage systems such as Azure Blob Storage or Azure Cosmos DB, or custom endpoints such as HTTP endpoints or Service Fabric applications.

5. **Message forwarding**: The IoT hub forwards the message to the specified destination(s) based on the routing rule evaluation. The message can be sent as-is or with modifications, depending on the configuration.

6. **Endpoint processing**: The destination endpoint(s) receive the message and perform the necessary processing or actions based on the specific requirements; for example, storing the data, triggering events, or integrating with other systems.

Routing in an IoT hub provides flexibility and scalability in handling large volumes of device data and distributing it to different services or applications based on predefined rules. It allows you to decouple device communication from backend processing and enables you to build complex IoT solutions with ease.

The following screenshot shows the various types of messages you can use to trigger a route:

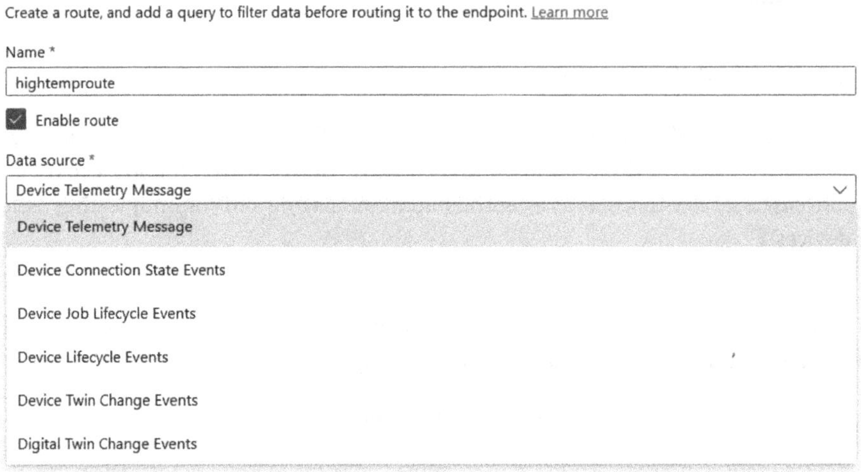

Figure 6.4 – Various types of data used to trigger a routing rule

Once you create a routing query, you should test it before you save it, as in the next screenshot:

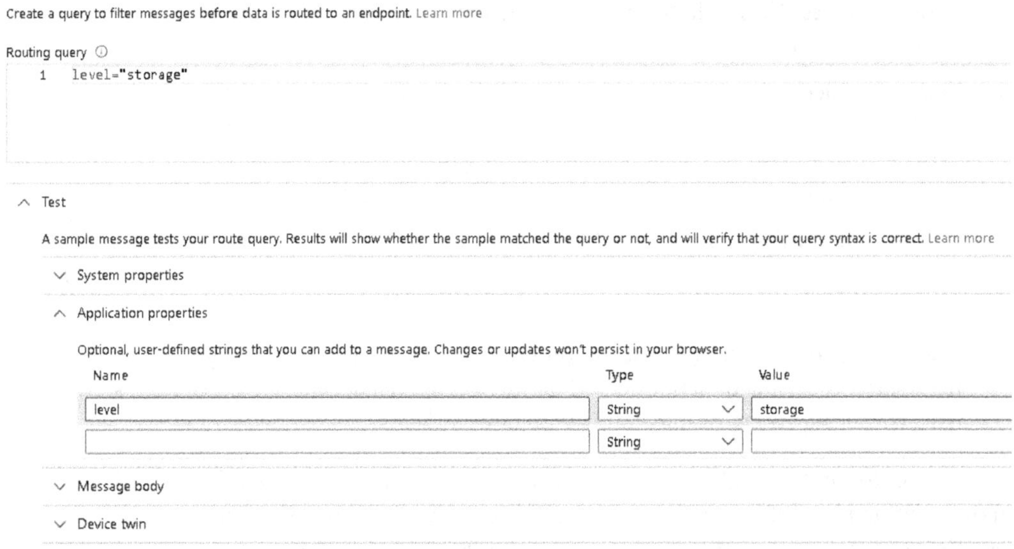

Figure 6.5 – Entering and testing a routing query

Once you save the route, you should verify that messages are being sent to the endpoint according to the query. In this case, messages should be showing up in the storage account if they are marked as `level` equal to `storage`.

Next, you will see how you can add enrichments to your messages.

Exploring message enrichment

Message enrichment in an IoT hub refers to the process of enhancing or augmenting incoming messages from IoT devices with additional information or metadata before they are routed or processed further. This enrichment can include adding contextual data, transforming the message format, or enriching the message payload with additional attributes.

IoT Hub provides a feature called **message enrichments** that allows you to configure rules to enrich messages as they pass through the hub. These rules can be defined based on specific criteria or conditions, and they enable you to modify messages by adding, updating, or deleting message properties or inserting custom data.

Here are a few common scenarios where message enrichment can be useful:

- **Adding metadata**: You can enrich messages with additional metadata such as device properties, location information, or timestamp information. This metadata can provide context to the message and help in downstream processing and analysis.

- **Device-to-cloud transformation**: Message enrichment can be used to transform the message format or structure from the device to a standardized format that is compatible with downstream systems or services. For example, you can convert the payload from a proprietary format to JSON or XML.

- **Data aggregation**: Enrichment rules can be defined to aggregate data from multiple messages or devices. This can involve combining or summarizing data to create a more comprehensive view of the system or to generate aggregated analytics.

- **Filtering or routing based on enriched properties**: Enriching messages allows you to add new properties based on specific conditions or calculations. You can then use these enriched properties to filter or route messages to different endpoints or processing pipelines based on the desired criteria.

To add message enrichment, select the **Message routing** in the left side menu of IoT Hub in the portal. Click on the **Enrich messages** tab on the **Message routing** page. As you can see in the following screenshot, the enrichment is associated with a particular endpoint. The value can be a constant string value or one obtained dynamically from variables exposed to you, such as $iothubname or $twin.tags.field. These values are then added as application properties to the messages:

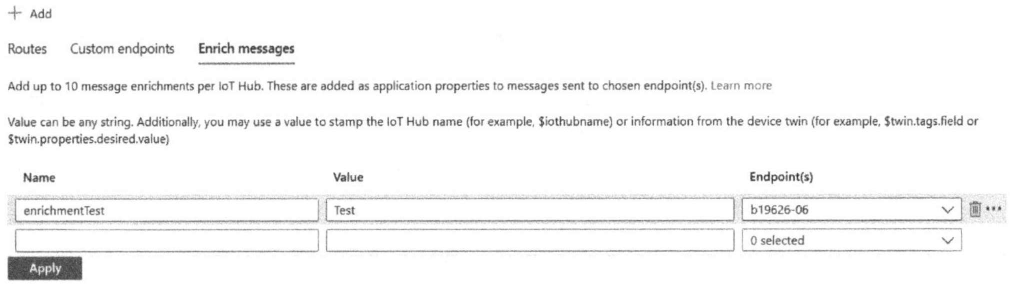

Figure 6.6 – Creating message enrichments

Message enrichment in an IoT hub provides a way to enhance information carried by IoT device messages, making it more meaningful and valuable for downstream processing, analysis, and integration with other systems. It allows you to customize and shape messages according to your specific requirements and enables richer and more flexible IoT solutions.

It is now time to put the knowledge you have gained into a hands-on lab activity.

Lab – using message enrichment and custom endpoints

Message routing within Azure IoT Hub serves as the conduit through which telemetry data from IoT devices is seamlessly channeled to an array of Azure services, including Blob Storage, Service Bus queues, Service Bus topics, and Event Hubs. Within the framework of IoT hubs, there exists a default, preconfigured endpoint that aligns seamlessly with Event Hubs. Furthermore, the system allows for the establishment of custom endpoints, coupled with the utilization of routing queries to precisely channel messages toward other Azure services. When a message arrives at the IoT hub, it is disseminated to all endpoints for which its routing queries are a match. In the event that a message fails to correspond with any of the defined routing queries, it is automatically directed toward the default endpoint.

In this lab, you will accomplish the following tasks:

- Establish an IoT hub and dispatch device messages to it
- Generate a storage account

- Create a custom endpoint for the storage account and configure message routing from the IoT hub to this endpoint

- Monitor device messages within the storage account blob

So, let's get started with the lab:

1. If you do not already have an IoT hub created, create one.

2. Under the **Device Management** section of IoT Hub, create a device.

3. Once created, copy the primary connection string of the device.

4. In the lab folder you downloaded from GitHub, go to the HubRoutingSample folder under the Lab6 folder and open a cmd console. (If you have not downloaded the labs, see the *Downloading the example code files* section in the *Preface* of this book).

5. Issue the dotnet restore command:

```
dotnet restore
```

6. Once restored, run the simulator by issuing the following command:

```
dotnet run --PrimaryConnectionString
<myDevicePrimaryConnectionString>
```

Make sure you provide your connection string and surround it with double quotes.

7. Once the simulator is running, verify that the IoT hub is receiving messages. If you want to download and install **Azure IoT Explorer** to view the messages, please refer to this source: https://learn.microsoft.com/en-us/azure/iot/howto-use-iot-explorer.

8. In the Azure portal, create a storage account in the same resource group as the IoT hub. For redundancy, select **Locally Redundant Storage (LRS)**.

9. Once created, locate the **Containers menu** item under the **Data storage** section of the storage account's menu. Create a container to store messages in.

10. To set up the routing, go back to the IoT hub in the Azure portal. Under the **Hub settings** section, select **Message routing** and then select **Add**.

11. Select a **Storage Endpoint type** value, enter an **Endpoint name** value, and select the storage container you created earlier. Select **JSON Encoding**.

12. Notice the **File name format** type. Then, click the **Create + next** button.

13. This is where you will set up and test your routing query. Name your route. For the **Data source** field, choose **Device Telemetry Message**.

14. Under **Routing query**, enter the following:

```
level="storage"
```

15. Under the **Test** option, add a **String** application property named `level` with a value of `storage`:

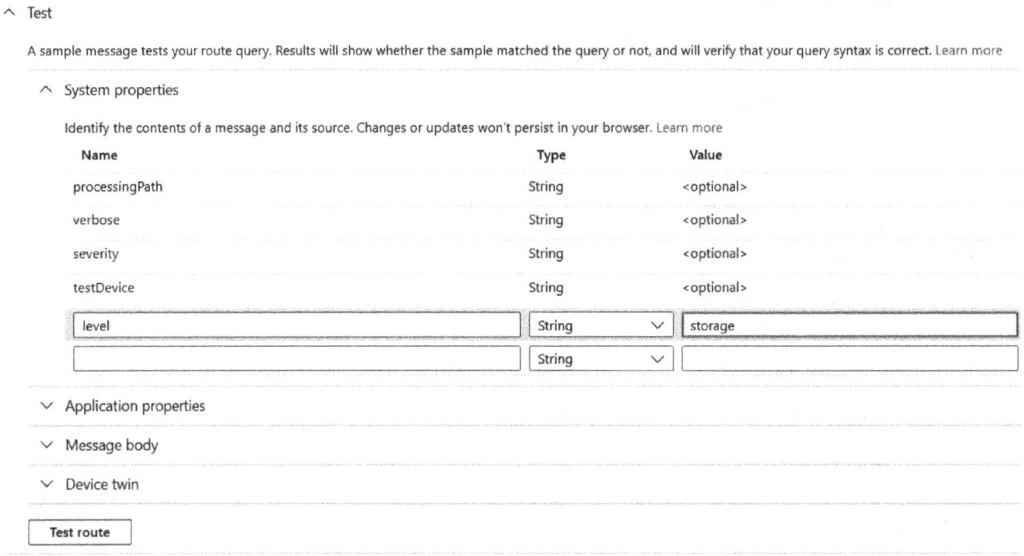

∧ Test

A sample message tests your route query. Results will show whether the sample matched the query or not, and will verify that your query syntax is correct. Learn more

   ∧ System properties

     Identify the contents of a message and its source. Changes or updates won't persist in your browser. Learn more

Name	Type	Value
processingPath	String	<optional>
verbose	String	<optional>
severity	String	<optional>
testDevice	String	<optional>
level	String ∨	storage
	String ∨	

   ∨ Application properties

   ∨ Message body

   ∨ Device twin

   Test route

Figure 6.7 – Adding a String application property

16. Click the **Test route** button, and if you entered everything correctly, it should succeed. Make sure it succeeds before moving on.

17. Once saved, select the **Enrich messages** tab. Create a name of `IoT Hub` and a value of `$iothubname`, and then select the endpoint you just created.

18. To test your route and enrichment, run the simulator if it is not already running. You should see messages starting to accumulate in the storage container. Open one of the files and observe the messages.

Now that you have seen how to implement message routing, let's review what we have learned in this chapter and where we are going next.

Summary

In this chapter, we dove into the world of Azure IoT Hub routing, exploring its capabilities, configuration options, and benefits. Azure IoT Hub routing is a powerful feature that allows users to efficiently route and process data generated by IoT devices, enabling seamless integration with downstream services and applications. IoT Hub routing acts as a central hub for data ingestion from multiple devices, providing a scalable and reliable platform for managing IoT data at scale.

We investigated the various routing capabilities offered by Azure IoT Hub. We discussed message routing, which allows users to define routing rules based on message properties, ensuring that data is sent to the appropriate endpoints. We also explored the flexibility of message routing, including the ability to route messages based on device properties, custom message properties, or even message content. Then, we covered message enrichment, whereby users can enrich messages with additional information during the routing process.

Message routing plays a pivotal role in the realm of data-driven systems, serving as the linchpin that ensures the efficient, reliable, and targeted flow of information. Its importance lies in its ability to orchestrate the journey of data from source to destination, often involving complex and diverse networks of interconnected devices and services. By intelligently directing messages based on predefined criteria or dynamically changing conditions, message routing optimizes data transmission, reduces latency, and enhances scalability. It enables the right information to reach the right destination at the right time, facilitating timely decision-making, process automation, and seamless integration between different components of a system.

In the next chapter, we turn our attention to ASA, which is a stream processing engine that is intended to analyze and process large volumes of streaming data with sub-millisecond latencies. It includes a robust SQL-like query language that you can use to query your streaming data to detect anomalies, aggregate data over different time windows, and perform geospatial analysis for moving sensors. You will learn how to incorporate this vital tool as part of your stream processing.

7

Exploring Azure Stream Analytics

Azure Stream Analytics (ASA) is a stream processing engine that is intended to analyze and process large volumes of streaming data with sub-millisecond latencies. It includes a robust SQL-like query language that you can use to query your streaming data to detect anomalies, aggregate data over different time windows, and perform geospatial analysis for moving sensors.

In this chapter, you will learn how to incorporate this vital tool as part of your stream processing. To fully comprehend how Stream Analytics processes data, we will dissect its inputs and outputs. We'll uncover how data is ingested from an IoT device and transformed into actionable insights, allowing organizations to react swiftly to changing conditions.

As we venture further into the world of Stream Analytics, we'll acquaint ourselves with the Stream Analytics query language. This language serves as the backbone of the system, providing a robust toolkit for filtering, aggregating, and transforming data on the fly.

But our journey won't stop at theory alone. We'll also learn how to put this knowledge into practice by exploring the intricacies of running and monitoring Stream Analytics jobs. Understanding the operational aspects of Stream Analytics is crucial for ensuring the reliability and performance of your real-time data processing pipeline.

Finally, we'll cap off our exploration with a hands-on laboratory experience. In the *Lab – detecting anomalies with ASA* section, we will apply the concepts and skills we've acquired throughout this chapter to detect anomalies in real-time streaming data. This practical exercise will help solidify your understanding of Stream Analytics and empower you to harness its capabilities for your own data-driven endeavors.

In particular, we will look at the following topics:

- Stream analytics use cases
- Inputs and outputs

- Stream Analytics query language

- Running and monitoring jobs

- Lab – detecting anomalies with ASA

Stream analytics use cases

Stream analytics, also known as **real-time analytics** or **real-time data processing**, involves analyzing and deriving insights from data that is generated continuously and in real time. Here are some common use cases for stream analytics:

- **Fraud detection**: Stream analytics can be used to detect fraudulent activities in real time. By continuously analyzing incoming data from various sources, such as financial transactions or user behavior patterns, patterns indicative of fraud can be identified and appropriate actions can be taken immediately.

- **IoT data processing**: IoT generates vast amounts of data from connected devices. Stream analytics can process and analyze this data in real time, enabling real-time monitoring, anomaly detection, predictive maintenance, and operational optimization.

- **Network monitoring and anomaly detection**: Stream analytics can be used to monitor network traffic and detect anomalies or suspicious activities in real time. This helps in identifying network intrusions, network performance issues, or any unusual behavior that could potentially indicate a security breach.

- **Supply chain optimization**: Real-time analytics can provide insights into supply chain processes by analyzing data from multiple sources such as inventory levels, customer demand, logistics, and weather conditions. This enables organizations to optimize their supply chain operations, make real-time adjustments, and improve efficiency.

- **Social media analytics**: Streaming data from social media platforms can be analyzed in real time to understand trends, **sentiment analysis (SA)**, customer feedback, and brand reputation. This enables businesses to respond quickly to customer needs, engage in real-time marketing, and make informed decisions based on the insights gained.

- **Predictive maintenance**: By continuously analyzing sensor data from equipment and machinery, stream analytics can detect patterns or anomalies that indicate potential failures. This enables proactive maintenance and reduces downtime by scheduling maintenance activities before critical failures occur.

- **Real-time personalization**: Stream analytics can be used to personalize user experiences in real time by analyzing user behavior data. This enables organizations to deliver targeted content, recommendations, and advertisements based on individual preferences and interactions.

- **Energy management**: Real-time analysis of data from smart meters, weather forecasts, and other sources can help optimize energy consumption, identify energy waste, and improve overall energy efficiency in buildings or industrial processes.

These are just a few examples of the many use cases for stream analytics. The versatility of this technology allows for its application in various industries, including finance, healthcare, retail, transportation, and more.

In the next section, you will look at getting data into and out of an ASA job.

Inputs and outputs

Before we can look at inputs and outputs, we need to create an ASA job. In your resource group, select **Add a resource** and search for `azure stream analytics`:

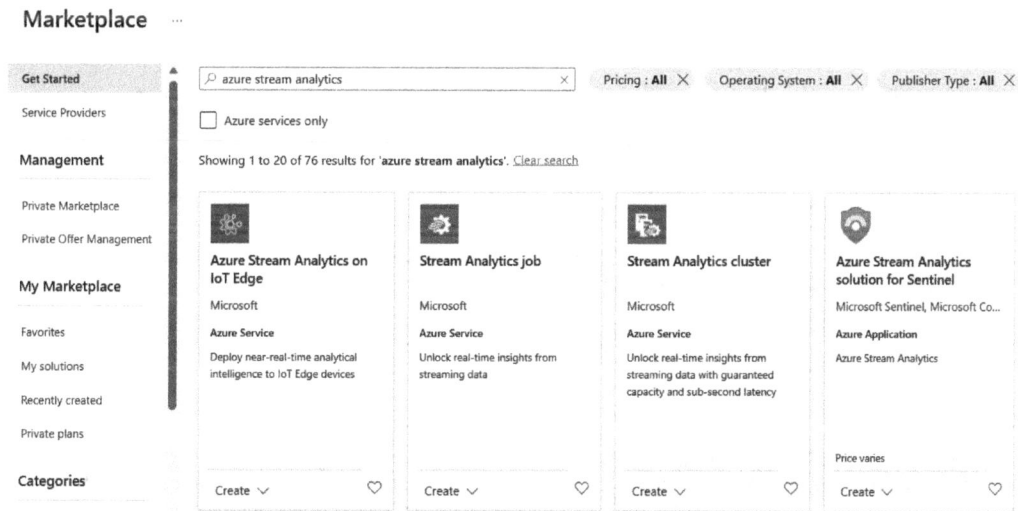

Figure 7.1 – Creating a cloud ASA job

While creating a Stream Analytics job service, you will see you can host it in the cloud or on the edge. For now, we are going to host it in the cloud. You also need to set the number of **streaming units** (**SUs**). When you create a Stream Analytics job in Azure, you can configure the number of SUs to allocate to that job. The number of **Azure Scale Units** (**ASUs**) determines the job's processing capacity, and it affects how much data the job can process in real time. For demo purposes, we can just use the default of three units:

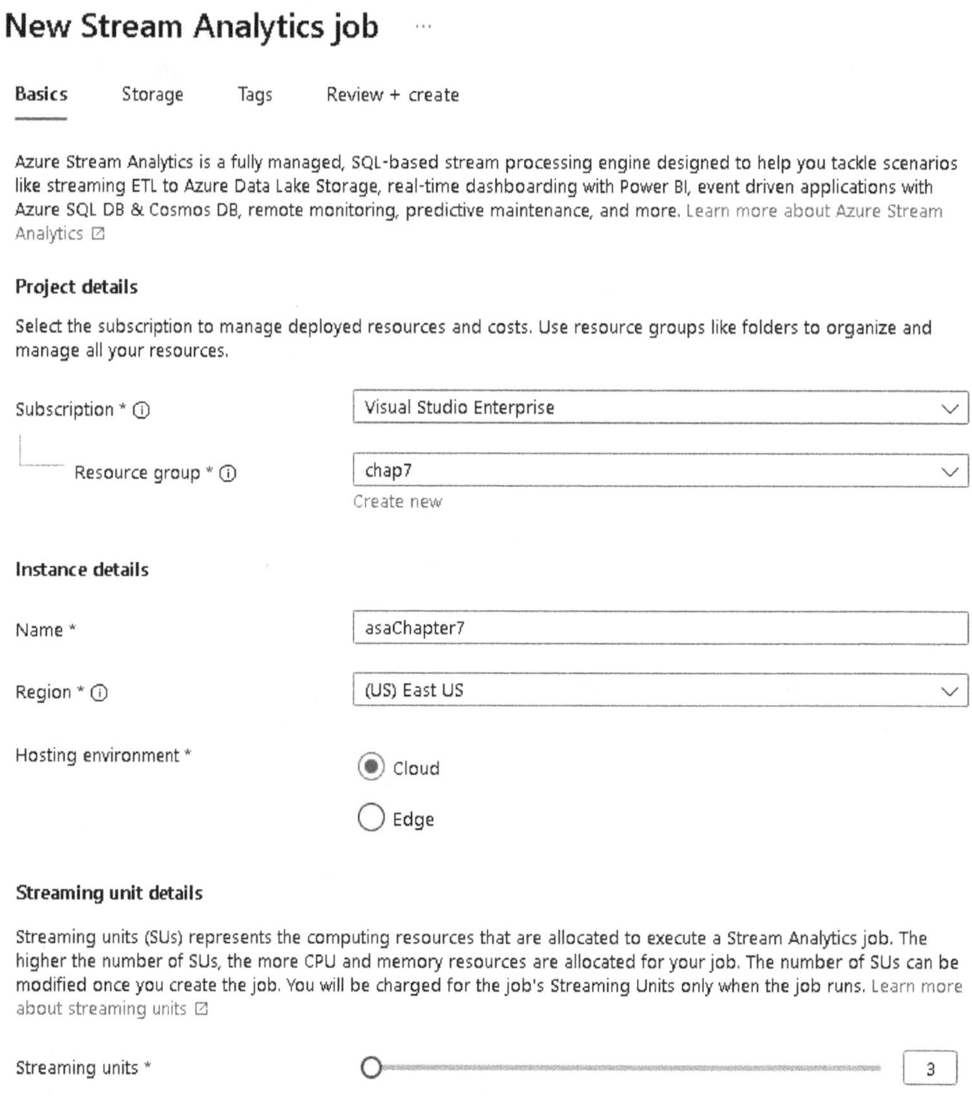

New Stream Analytics job ...

Basics   Storage   Tags   Review + create

Azure Stream Analytics is a fully managed, SQL-based stream processing engine designed to help you tackle scenarios like streaming ETL to Azure Data Lake Storage, real-time dashboarding with Power BI, event driven applications with Azure SQL DB & Cosmos DB, remote monitoring, predictive maintenance, and more. Learn more about Azure Stream Analytics ☑

Project details

Select the subscription to manage deployed resources and costs. Use resource groups like folders to organize and manage all your resources.

Subscription * ⓘ	Visual Studio Enterprise ⌄
Resource group * ⓘ	chap7 ⌄
	Create new

Instance details

Name *	asaChapter7
Region * ⓘ	(US) East US ⌄
Hosting environment *	● Cloud
	○ Edge

Streaming unit details

Streaming units (SUs) represents the computing resources that are allocated to execute a Stream Analytics job. The higher the number of SUs, the more CPU and memory resources are allocated for your job. The number of SUs can be modified once you create the job. You will be charged for the job's Streaming Units only when the job runs. Learn more about streaming units ☑

Streaming units *	○————————————— 3

Figure 7.2 – Setting up a Stream Analytics job

When setting up a Stream Analytics job, you first need to define an input. The two types of inputs are **stream inputs** and **reference inputs**:

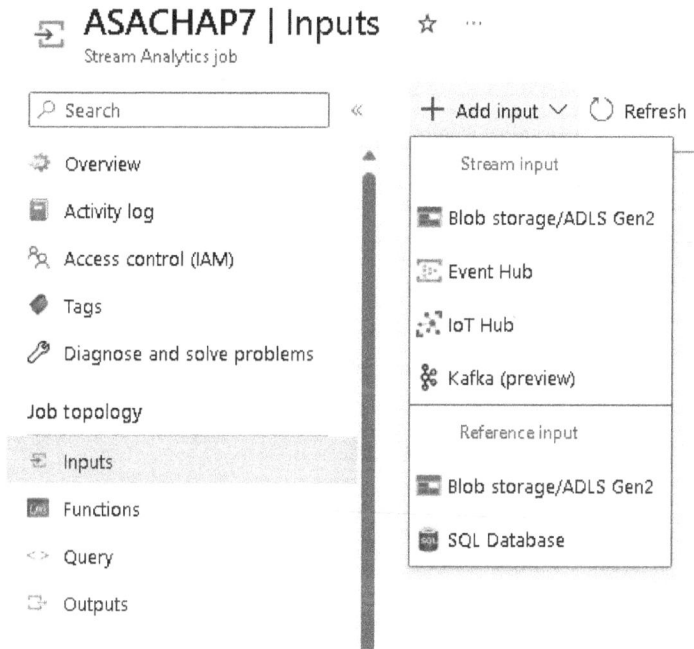

Figure 7.3 – Setting up inputs

The reference input is there to help augment the stream data coming in. You can use reference data to perform dimensional analysis, for example, converting location data to buildings or combining it with operational data such as truck information.

Overall, reference data in ASA enhances your ability to analyze and gain insights from streaming data by enriching it with relevant information, performing lookup operations, enabling dimensional analysis, and supporting dynamic updates.

The stream input data can come from a variety of sources, including Blob torage, event hubs, IoT hubs (most relevant for us), and Kafka as it is the most common and easy to integrate, as you can see in *Figure 7.3* on the **Add input** dropdown.

Along with inputs and outputs (more on outputs in the following paragraphs), you can also use functions during stream processing. By combining Azure Functions with ASA, you can extend the processing capabilities and incorporate custom logic, data transformations, advanced analytics, and integrations into your streaming data workflows. It provides a powerful mechanism for real-time data processing, enrichment, and integration with various systems and services in the Azure ecosystem.

In the following screenshot, an IoT hub is being set up as the input to the Stream Analytics job:

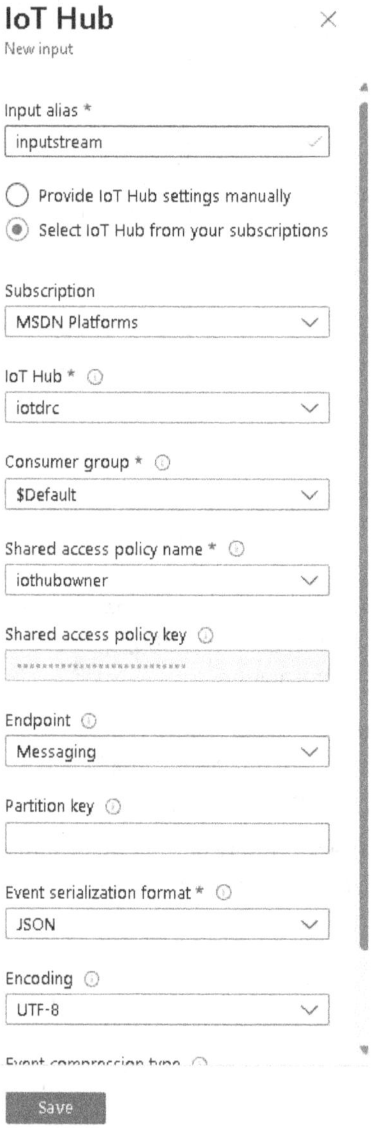

Figure 7.4 – Setting up an IoT hub as input data

Once the input is added, the next step is to add an output. As you can see, there are many output types you can easily set up in ASA. These depend on whether you are going to process the data further, store the data, or visualize the data:

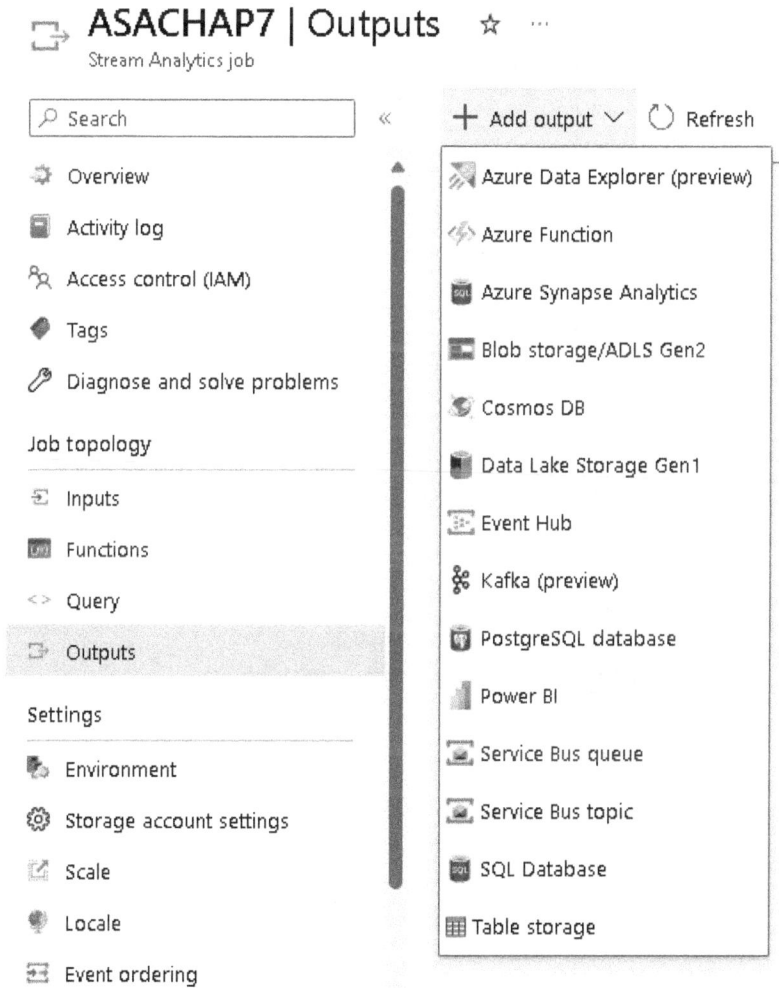

Figure 7.5 – Adding an output to the Stream Analytics job

The following screenshot shows setting up an Azure table (choose the **Table storage** option) for an output:

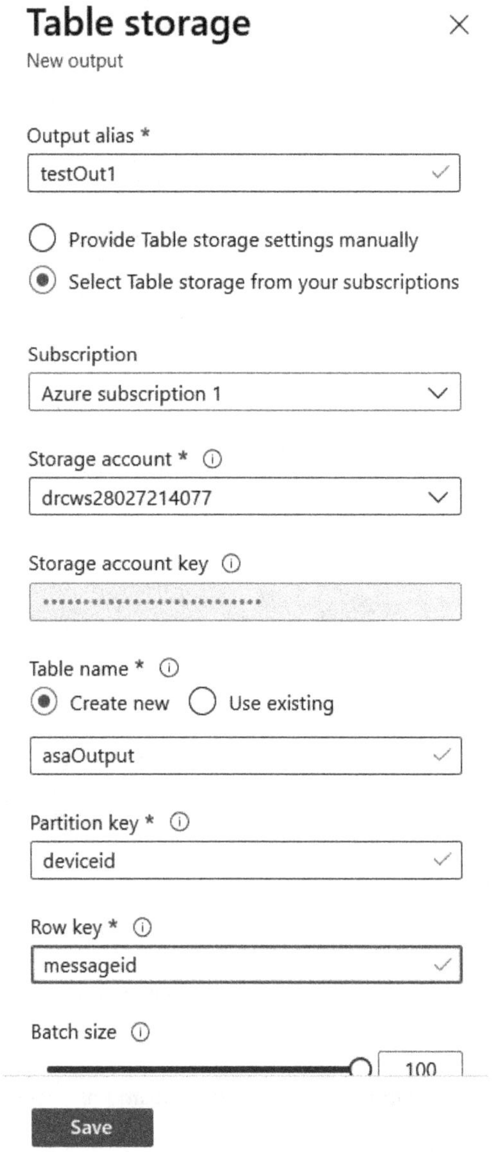

Table storage   ✕

New output

Output alias *

| testOut1   ✓ |

◯ Provide Table storage settings manually

◉ Select Table storage from your subscriptions

Subscription

| Azure subscription 1   ⌄ |

Storage account * ⓘ

| drcws28027214077   ⌄ |

Storage account key ⓘ

| •••••••••••••••••••••••••••• |

Table name * ⓘ
◉ Create new   ◯ Use existing

| asaOutput   ✓ |

Partition key * ⓘ

| deviceid   ✓ |

Row key * ⓘ

| messageid   ✓ |

Batch size ⓘ

[─────────────────────◯] | 100 |

[**Save**]

Figure 7.6 – Setting up an Azure table for streaming output

When setting up Azure tabular storage, you need to indicate the table name, keys, and batch size.

After setting up your inputs and outputs, you then set up the query. If you are used to working with streaming **Structured Query Language (SQL)**, you are in luck. ASA uses a variant of SQL, known as **Azure Streaming Query Language (ASQL)**, to define queries and perform real-time data processing on streaming data. ASQL provides a familiar SQL-like syntax with additional constructs for handling streaming data. Let's explore more about query language in the upcoming section.

Stream Analytics query language

ASQL simplifies the process of defining and executing real-time data processing logic on streaming data. Its SQL-like syntax, along with stream-specific constructs, enables developers and data engineers to perform complex analytics, transformations, and aggregations on streaming data with ease.

Here are some key features and concepts of ASQL:

- **Input and output sources**: ASQL allows you to define input sources, such as event hubs or IoT hubs, from which streaming data is ingested. You can specify the schema of the incoming data and define output sinks, such as Azure Blob Storage or Azure SQL Database, to store or forward the processed data.

- **Query structure**: ASQL queries typically follow a structure similar to SQL SELECT statements. However, ASQL queries operate on an infinite, unbounded stream of data, rather than discrete tables. The data is processed in an event-driven and continuous manner.

- **Windowing**: ASQL supports various windowing functions to define time-based or count-based windows over the streaming data. Windowing allows you to perform calculations or aggregations over specific time intervals or event counts, enabling real-time analytics.

- **Time management**: ASQL provides functions to handle event time and process time. Event time represents the time when an event occurred, while process time represents the time when the event is processed by Stream Analytics. You can use these functions for event time synchronization, handling out-of-order events, and managing time windows.

- **Joins and aggregations**: ASQL supports join operations, allowing you to correlate data from multiple input streams based on common fields or conditions. You can also perform aggregations such as COUNT, SUM, AVG, MAX, and MIN on the streaming data, enabling real-time analytics and calculations.

- **User-defined functions (UDFs)**: ASQL allows you to define custom UDFs using JavaScript. UDFs enable you to encapsulate complex logic, reuse code, and perform custom calculations or transformations on the streaming data.

- **Output operations**: ASQL provides output operations to define how the processed data should be written to output sinks. You can specify the output format, schema mapping, partitioning, and other properties for writing the results to storage or forwarding to other systems.

ASQL simplifies the process of defining and executing real-time data processing logic on streaming data.

ASA supports a wide range of queries to perform real-time data processing and analytics on streaming data. Here are some common query examples:

- The most basic query is to pass data from the input to the output:

```
SELECT
    *
INTO
    [YourOutputAlias]
FROM
    [YourInputAlias]
```

- Here's a simple filtering query:

```
SELECT *
FROM InputStream
WHERE temperature > 30
```

This query selects all events from InputStream where the temperature is greater than 30.

- This query groups events from InputStream by region within a 10-second tumbling window and calculates the count of events for each region within the window:

```
Aggregation Query:
SELECT region, COUNT(*) AS count
FROM InputStream
GROUP BY region, TumblingWindow(second, 10)
```

- This query joins data coming in the stream with reference data used to find the location of the device:

```
SELECT A.deviceId, A.temperature, B.location
FROM TemperatureStream A
JOIN LocationStream B
ON A.deviceId = B.deviceId
```

These are just a few examples of the queries you can write in ASA. The queries can be customized based on your specific use case, including data transformations, complex aggregations, filtering, joining multiple streams, and writing the results to various output sinks.

The following screenshot shows entering a query using the portal. If you have a sample of the input data, you can test the query and view the output:

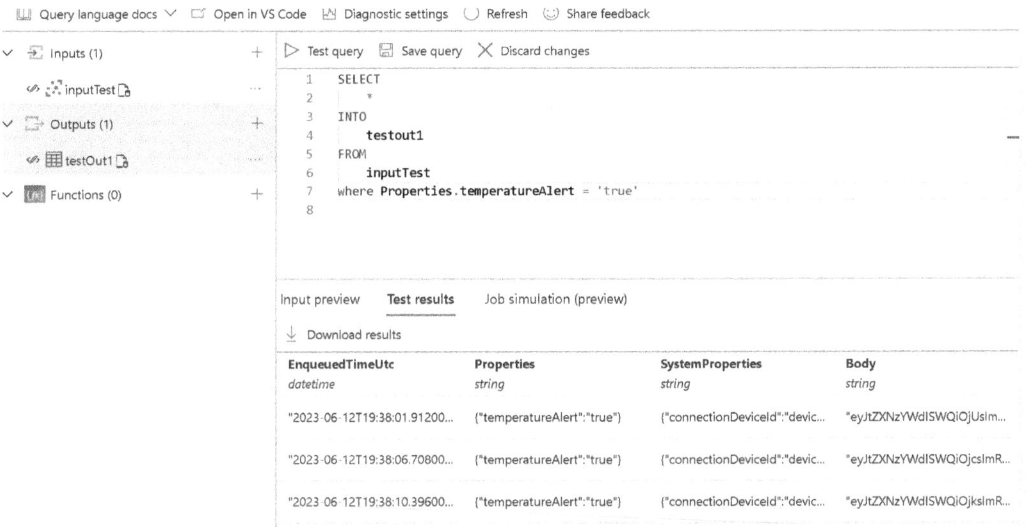

Figure 7.7 – Writing and testing queries in ASA

The query in *Figure 7.7* is filtering out data except when there is a temperature alert.

Now that you can set up ASA queries, it is time to run an ASA job. In the next section, you will see how to run and monitor jobs.

Running and monitoring jobs

After creating and testing a job, the next step is running and monitoring it to ensure it is processing data as expected and promptly detect any issues or anomalies.

Before running a job, you should set up the diagnostic settings by going to **Diagnostic settings** on the left side menu. The following screenshot demonstrates setting up execution logs saved to a storage account:

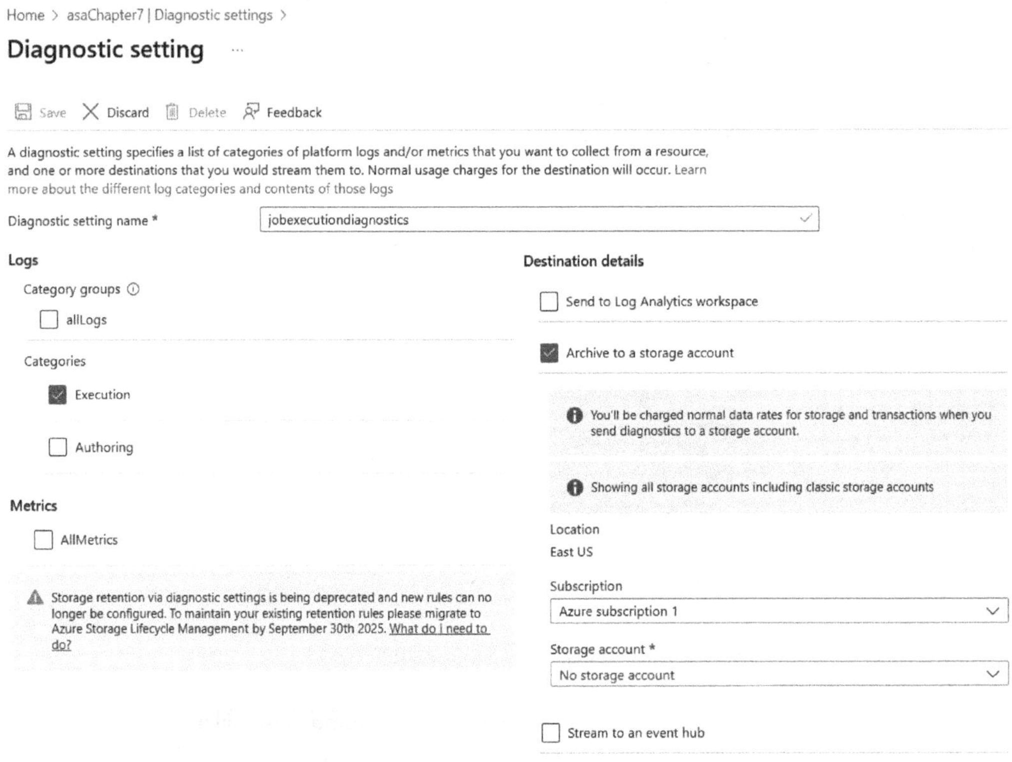

Home > asaChapter7 | Diagnostic settings >

Diagnostic setting   ...

🖫 Save   ✕ Discard   🗑 Delete   ⧉ Feedback

A diagnostic setting specifies a list of categories of platform logs and/or metrics that you want to collect from a resource, and one or more destinations that you would stream them to. Normal usage charges for the destination will occur. Learn more about the different log categories and contents of those logs

Diagnostic setting name * | jobexecutiondiagnostics   ✓ |

Logs   **Destination details**

Category groups ⓘ   ☐ Send to Log Analytics workspace
  ☐ allLogs

Categories   ☑ Archive to a storage account

  ☑ Execution   ❶ You'll be charged normal data rates for storage and transactions when you
                                                                  send diagnostics to a storage account.
  ☐ Authoring

                                                               ❶ Showing all storage accounts including classic storage accounts
Metrics
                                                            Location
  ☐ AllMetrics   East US

  ⚠ Storage retention via diagnostic settings is being deprecated and new rules can no   Subscription
     longer be configured. To maintain your existing retention rules please migrate to | Azure subscription 1   ∨ |
     Azure Storage Lifecycle Management by September 30th 2025. What do I need to
     do?   Storage account *
                                                            | No storage account   ∨ |

                                                            ☐ Stream to an event hub

Figure 7.8 – Setting up diagnostic logs

In this case, we are looking at execution logs and retaining them for 10 days.

ASA also allows you to set up alerts based on specific conditions or metrics. By creating alerts, you can receive notifications when certain thresholds or conditions are met. In the Azure portal, go to your Streaming Analytics service, click on **Alerts** on the left-hand menu, and configure the alert rules based on your requirements.

Once the job is created, on the job overview page, click on the **Start job** button to start the job. The job will begin processing data based on the specified query logic:

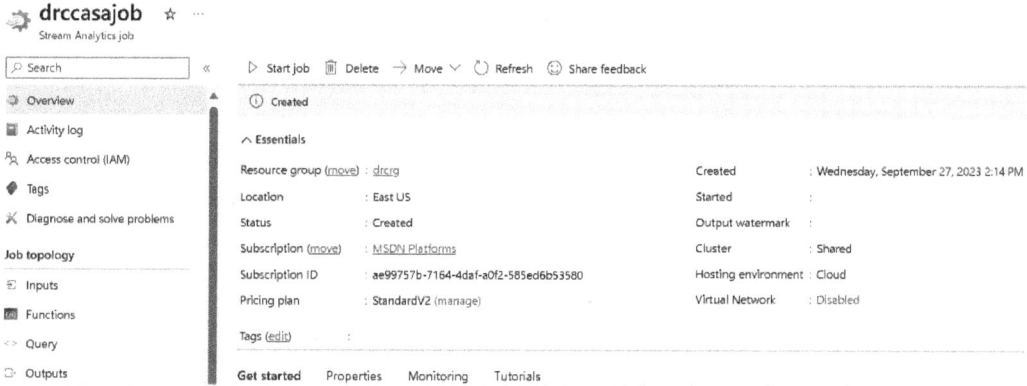

Figure 7.9 – Starting a job

After starting the job, you can monitor its status and health. On the job details page, you will see the job status, such as **Running**, **Starting**, or **Stopping**. You can also view other details such as input and output events, ingress and egress rates, and any error messages or warnings, as shown in the following screenshot:

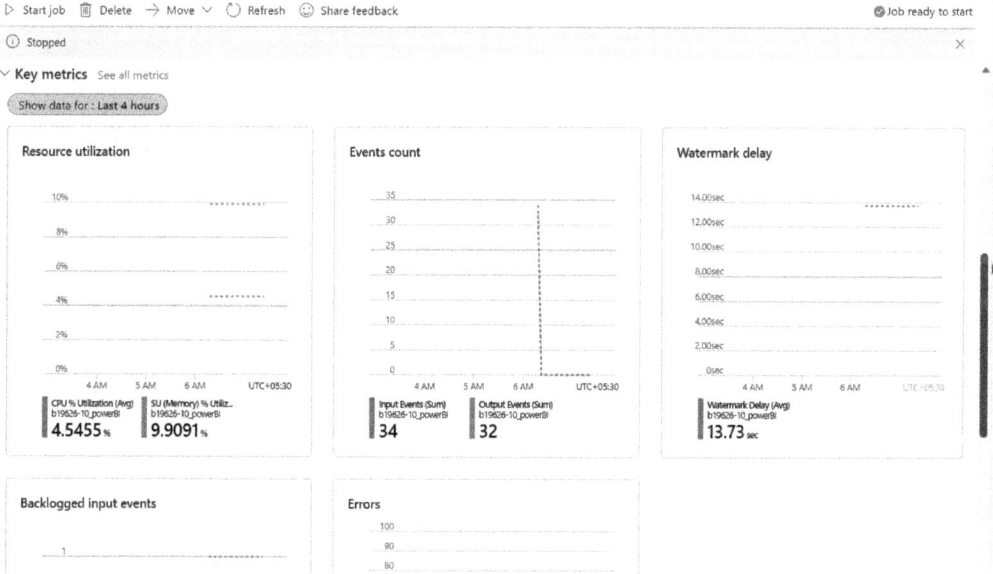

Figure 7.10 – Monitoring a running job

Now that you have seen how to set up, run, and monitor ASA jobs, it is time to gain some hands-on experience in the following lab.

Lab – detecting anomalies with ASA

One of the most common use cases of ASA is to detect anomalies. In this lab, you will see how it is done:

1. Create an IoT hub with a device named `device1`.

2. Go to the device simulator at `https://azure-samples.github.io/raspberry-pi-web-simulator` and enter your device connection string.

3. Start the simulator and make sure your IoT hub is getting messages. You can use the following CLI command to view messages:

    ```
    az iot hub monitor-events --hub-name
    ```

4. If it asks you to install dependencies, choose **Yes**.

5. You can stop the simulator while you set up the ASA job.

6. Create a blob storage account to save the results of the job. Create a blob container named `logs`.

7. In the same resource group, select **Create a resource** and search for `stream analytics job`.

    Create a Stream Analytics job using the default values.

8. Once created, open the Stream Analytics job in the Azure portal and select **Inputs** on the left-hand menu. Add an IoT hub input. Select the IoT hub you created earlier. Make sure you test the input:

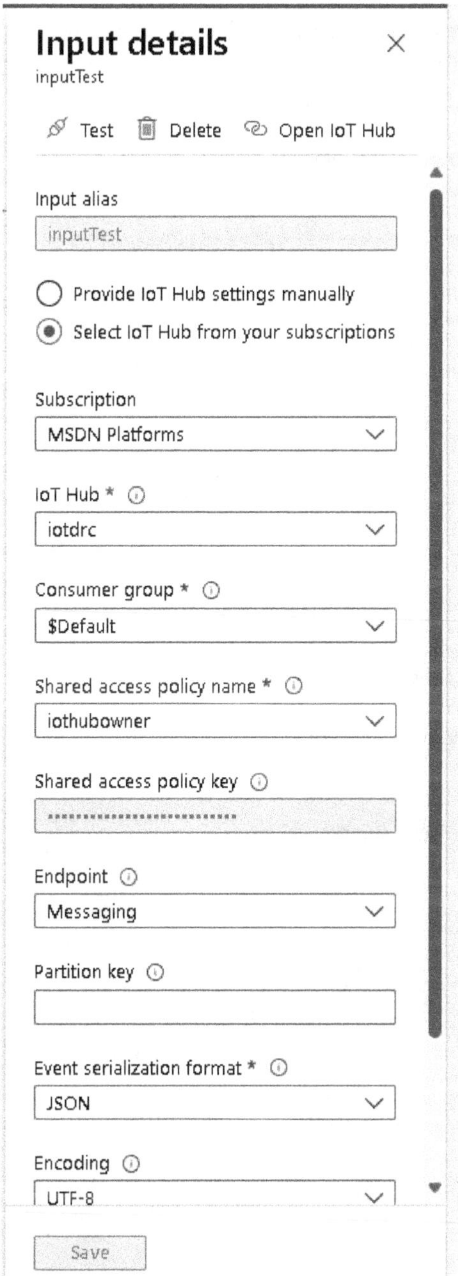

Input details

inputTest

🖉 Test   🗑 Delete   ⚭ Open IoT Hub

Input alias

inputTest

○ Provide IoT Hub settings manually

◉ Select IoT Hub from your subscriptions

Subscription

MSDN Platforms

IoT Hub * ⓘ

iotdrc

Consumer group * ⓘ

$Default

Shared access policy name * ⓘ

iothubowner

Shared access policy key ⓘ

•••••••••••••••••••••••••••••

Endpoint ⓘ

Messaging

Partition key ⓘ

Event serialization format * ⓘ

JSON

Encoding ⓘ

UTF-8

Save

Figure 7.11 – Adding a job input

9. Select **Outputs** on the left-hand menu. Create a blob storage output pointing to the `logs` container you created earlier. For **Path pattern**, use `logs/{datetime:MM}/{datetime:dd}`. Save and wait a few minutes to test the output:

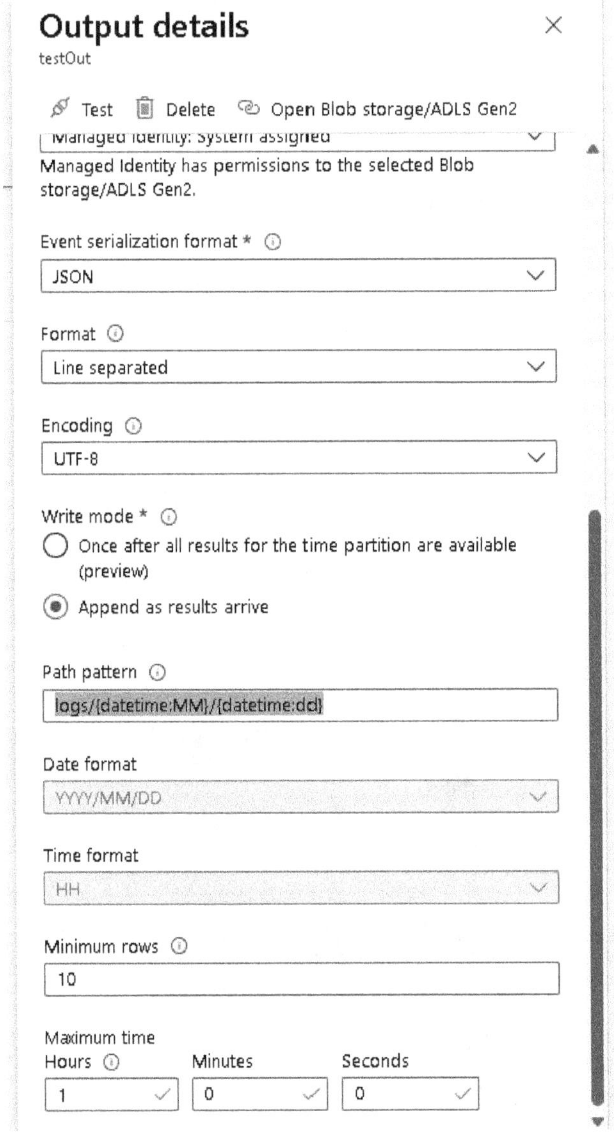

Figure 7.12 – Adding an output

10. Open the **Query** tab from the left-hand menu and enter the following query:

```
SELECT *
INTO
Testoutput
FROM
InputTest
where temperature > 27 and humidity > 70
```

Remember to replace `testOut` and `inputTest` with the name of your input and output.

11. You can test the query by uploading the `inputTest.json` test file in the `Lab7` folder in the GitHub downloads.

12. Start the virtual simulator and the ASA job. Wait about 5 minutes for everything to start.

After a few minutes, open the output blob file in the storage account. You should see records being written to the file:

logs/06/18/0_f4dc9c23b6644deb8cf19803c7041bab_1.json ···   ×
Blob

💾 Save   ✕ Discard   ↓ Download   ⟳ Refresh | 🗑 Delete

Overview   Versions   Snapshots   **Edit**   Generate SAS

```
 1  {"messageId":16,"deviceId":"Raspberry Pi Web Client","temperature":28.361914331021342,"humidity":73.7275672489355,"Even
 2  {"messageId":19,"deviceId":"Raspberry Pi Web Client","temperature":31.02342887167635,"humidity":76.58565925529112,"Eve
 3  {"messageId":26,"deviceId":"Raspberry Pi Web Client","temperature":28.29803683000736,"humidity":76.02612537463503,"Even
 4  {"messageId":35,"deviceId":"Raspberry Pi Web Client","temperature":29.107773199913403,"humidity":74.17421248405319,"Eve
 5  {"messageId":36,"deviceId":"Raspberry Pi Web Client","temperature":29.059052809081464,"humidity":72.11207339265329,"Eve
 6  {"messageId":39,"deviceId":"Raspberry Pi Web Client","temperature":31.244906036646015,"humidity":75.25059742726574,"Eve
 7  {"messageId":42,"deviceId":"Raspberry Pi Web Client","temperature":31.002042483444107,"humidity":77.29225623959422,"Eve
 8  {"messageId":46,"deviceId":"Raspberry Pi Web Client","temperature":30.678932929166702,"humidity":77.14200727098479,"Eve
 9  {"messageId":52,"deviceId":"Raspberry Pi Web Client","temperature":27.598215068346825,"humidity":74.06149769013838,"Eve
10  {"messageId":66,"deviceId":"Raspberry Pi Web Client","temperature":30.115241582715274,"humidity":72.6368167820722,"Even
```

Figure 7.13 – Viewing the output records

13. When finished, stop the job and the simulator.

While this example used static values to identify anomalies, ASA easily allows for the integration of **machine learning** (ML) to determine anomalies and detect fraud. If that is something you want to implement, I would strongly recommend searching the internet for more examples.

Summary

In this chapter, the focus was on ASA, a powerful stream processing engine designed to handle and analyze vast amounts of streaming data with minimal latency. The chapter began by introducing the fundamental concepts and features of ASA.

The chapter emphasized the significance of using the robust SQL-like query language offered by ASA to efficiently query streaming data. This query language enables users to detect anomalies in the data, aggregate data over different time windows, and perform geospatial analysis for moving sensors. By mastering this query language, you can leverage the full potential of ASA for your stream processing needs.

Additionally, the chapter covered essential aspects such as setting up inputs and outputs for ASA. Understanding how to configure and manage these data sources and destinations is crucial for effective stream processing.

The next chapter delves into the capabilities of Azure Data Explorer, a fully managed, high-performance analytics platform designed to handle large volumes of data in near real time. The chapter highlights the comprehensive toolbox offered by Azure Data Explorer, which provides end-to-end solutions for data ingestion, query execution, visualization, and management.

Azure Data Explorer empowers users to analyze a wide range of data types, including structured, semi-structured, and unstructured data, with a specific focus on time series analysis. By leveraging ML techniques, users can effortlessly extract valuable insights, identify patterns and trends, and even create accurate forecasting models. The platform's scalability, security features, robustness, and enterprise readiness make it a valuable tool for various use cases, including log analytics, time series analytics, IoT data processing, and general-purpose exploratory analytics.

8

Investigating IoT Data with Azure Data Explorer

In today's data-driven world, organizations are faced with an overwhelming amount of information flowing through their systems. This is especially true of many IoT and logging solutions. To stay ahead of the game, it is crucial to efficiently collect, store, analyze, and visualize this data to extract valuable insights. This is where **Azure Data Explorer** (**ADX**) comes into play, which is a lightning-fast data analytics service provided by Microsoft Azure.

In this chapter, we will dive into the world of ADX and explore its capabilities, its features, and how it can revolutionize the way you work with data. We will walk you through the essential concepts and a hands-on example that will help you understand how ADX is used. While we will focus on how it is used for analyzing IoT data, it can also be used with a vast amount of different data types and sources.

Here's an overview of what you can expect to learn in this chapter:

- What is Azure Data Explorer?
- Ingesting streaming data
- Time series analysis
- Visualizing data in Azure Data Explorer
- Lab – creating an Azure Data Explorer dashboard

What is Azure Data Explorer?

Azure Data Explorer (**ADX**) is an advanced and fully managed analytics platform specifically designed for processing vast amounts of data with incredible speed and efficiency. With its comprehensive set of tools and features, ADX offers a complete solution for data ingestion, querying, visualization, and management.

By analyzing diverse data types such as structured, semi-structured, and unstructured data across time series, ADX empowers users to extract valuable insights, identify patterns and trends, and even develop accurate forecasting models. It incorporates the power of machine learning, making complex analytical tasks remarkably simple.

Scalability, security, and robustness are at the core of ADX, making it an ideal choice for a wide range of applications. Whether you need to perform log analytics, time series analytics, IoT data analysis, or general-purpose exploratory analytics, ADX is a reliable and enterprise-ready solution.

ADX is built to handle high volumes of data and deliver near real-time analytics. It achieves exceptional query performance through its optimized indexing and columnar storage techniques, enabling users to explore and analyze massive datasets with remarkable speed.

ADX specializes in time series data analysis. It provides native support for ingesting, storing, and querying time-stamped data, making it ideal for analyzing data streams from IoT devices, log files, financial data, and more. Its powerful time-based functions and windowing capabilities enable users to perform advanced time series analytics effortlessly.

With ADX, users have the freedom to explore diverse types of data, including structured, semi-structured, and unstructured data. Its query language, **Kusto Query Language** (**KQL**), provides a rich set of functions and operators to manipulate and extract insights from data efficiently. This flexibility makes it suitable for a wide range of use cases and empowers data analysts to uncover valuable information from complex datasets.

ADX prioritizes data security and compliance. It integrates with Azure Active Directory for identity and access management and provides robust encryption mechanisms to safeguard data at rest and in transit. Additionally, it complies with various industry standards, including GDPR, HIPAA, ISO, and SOC, ensuring that sensitive data is handled in a secure and compliant manner.

ADX is a fully managed service provided by Microsoft Azure. This means that Microsoft takes care of the underlying infrastructure, including maintenance, updates, and scaling, allowing users to focus on analyzing data and deriving insights rather than managing the underlying infrastructure.

To use ADX, you need to create it as a resource in your Azure subscription. To create an instance of ADX in the Azure portal, click the **Create a resource** link in the left-hand menu of the portal. Search for and select **Azure Data Explorer**. You will be asked to create an Azure Explorer cluster, as shown in *Figure 8.1*.

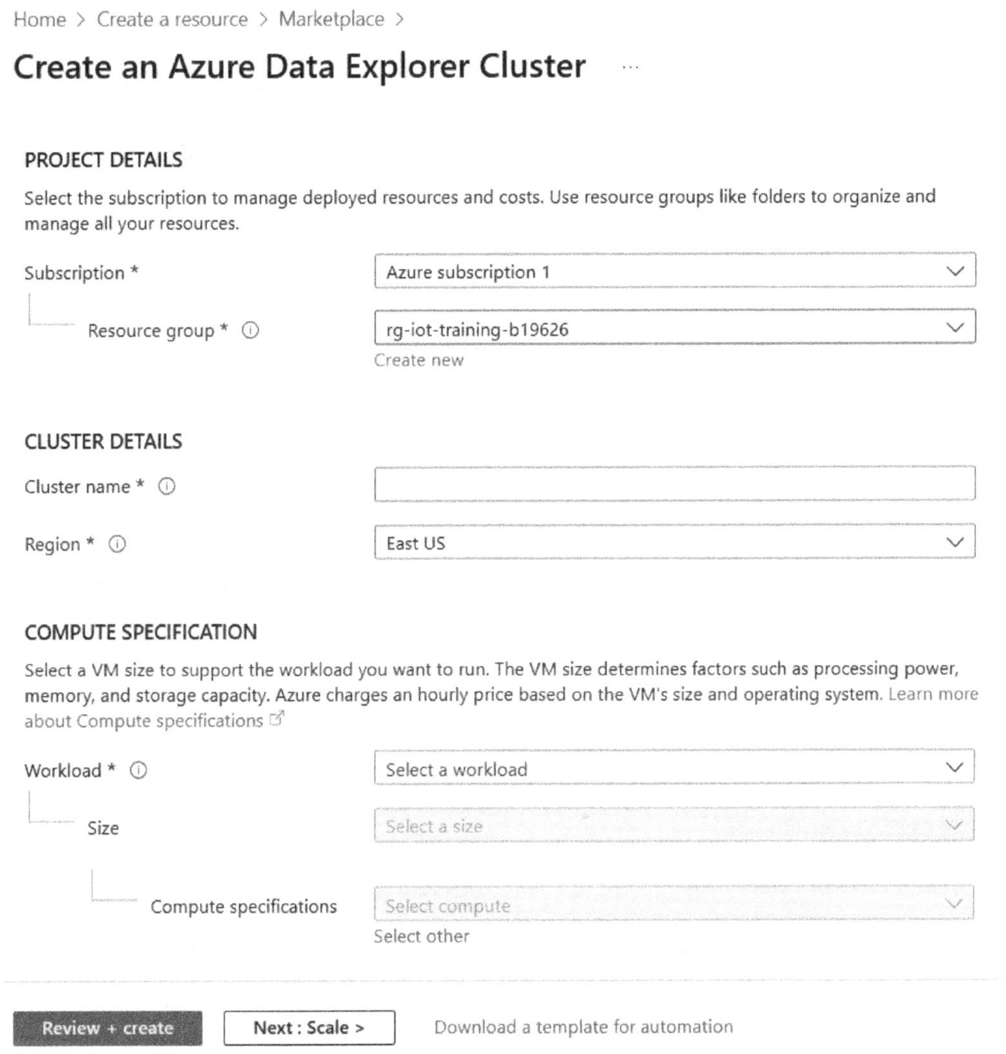

Home > Create a resource > Marketplace >

Create an Azure Data Explorer Cluster   ...

PROJECT DETAILS

Select the subscription to manage deployed resources and costs. Use resource groups like folders to organize and manage all your resources.

Subscription * | Azure subscription 1   ∨ |

└─── Resource group * ⓘ | rg-iot-training-b19626   ∨ |
                                    Create new

CLUSTER DETAILS

Cluster name * ⓘ | |

Region * ⓘ | East US   ∨ |

COMPUTE SPECIFICATION

Select a VM size to support the workload you want to run. The VM size determines factors such as processing power, memory, and storage capacity. Azure charges an hourly price based on the VM's size and operating system. Learn more about Compute specifications ☐

Workload * ⓘ | Select a workload   ∨ |

└─── Size | Select a size   ∨ |

      └─── Compute specifications | Select compute   ∨ |
                                    Select other

[Review + create]   [Next : Scale >]   Download a template for automation

Figure 8.1 – Creating an ADX cluster

For learning purposes, you will select **Dev/test workload**. You can also choose **Storage optimized** or **Compute optimized**. You can look through the rest of the tabs, but leave the defaults selected.

In a typical workflow of ADX, you will navigate through the following sequence of steps:

1. Initially, you will ingest your data into the system to bring it into the ADX environment.

2. Subsequently, you will proceed with analyzing the ingested data.

3. Once the analysis is performed, you will visualize the outcomes to gain meaningful insights.

4. Throughout the process, you may also utilize the data management functionalities as needed.

These interactions with ADX are conducted by engaging with the cluster, which serves as the central point of access. The resources and functionalities of the cluster can be accessed conveniently either through the web UI or by employing **Software Development Kits (SDKs)**. If you want to learn more about SDKs, this is a good place to start: `https://learn.microsoft.com/en-us/azure/data-explorer/kusto/api/`.

Now that we know what ADX is, in the next section, you will look at ingesting streaming data using ADX.

Ingesting streaming data

Although you can ingest a variety of data into Data Explorer for analysis, for our purposes we are going to look at ingesting IoT data streams.

Streaming ingestion is valuable for data loading when there is a need for minimal delay between the process of ingesting data and querying it. It is advisable to employ streaming ingestion in the following situations:

- When a latency of less than one second is necessary

- To enhance the operational processing of numerous tables where the flow of data into each table is relatively small (a few records per second), while the overall data ingestion volume is substantial (thousands of records per second)

- If the flow of data into each table is substantial (over 4 GB per hour), it is recommended to use batch ingestion

There are two types of ingestion supported in ADX: data connection and custom ingestion:

- **Data connection**: You can use this if the data is coming from Event Hubs, IoT Hub, and Event Grid. You can use streaming ingestion; however, it must be enabled on the cluster supporting your instance of ADX.

- **Custom ingestion**: This is used for loading data into the system from various external sources that are not directly supported by built-in connectors or plugins. It allows users to define and implement their own custom data ingestion pipelines to bring in data from diverse formats and locations.

  With custom ingestion, you can design and implement a data ingestion process tailored specifically to your data source, format, and transformation requirements. This flexibility enables you to extract data from virtually any source and load it into ADX for analysis and querying.

As mentioned previously, you need to enable streaming ingestion in the ADX cluster as shown here:

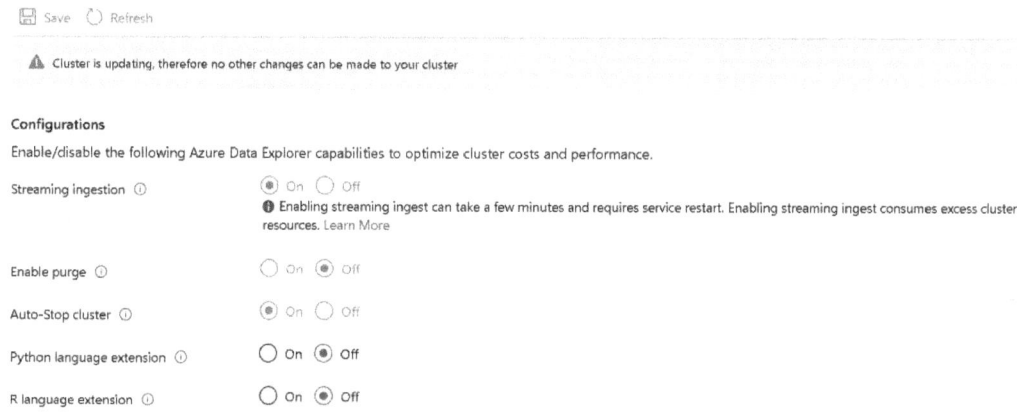

Figure 8.2 – Enabling streaming ingestion in the ADX cluster

Once you have enabled streaming ingestion, the next step is to create a table to collect the streaming ingestion data and set its related policy. If you know SQL, it will be a short learning curve to understand KQL. If you don't know SQL, there are many online resources you can use to learn the basics, including at https://www.w3schools.com/sql/.

The following code sets up a new table called Telemetry to capture the IoT input from a device:

```
.create table Telemetry ( messageId: int, deviceId: string,
temperature:decimal, humidity:decimal, temperatureAlert: string,
IotHubDeviceId: string, IotHubEnqueuedTime: datetime )
```

The next step is to define a mapping between the incoming message and the table. First, we enable streaming ingestion if not already enabled:

```
.alter table Telemetry policy streamingingestion enable
```

Then we define the mapping:

```
.create table Telemetry ingestion json mapping "JsonTelemetryMapping"
'[' '{"Column": "messageId", "Properties": {"Path": "$.messageId"}},'
'{"Column": "deviceId", "Properties": {"Path": "$.deviceId"}},'
'{"Column": "temperature", "Properties": {"Path": "$.temperature"}},'
'{"Column": "humidity", "Properties": {"Path": "$.humidity"}},'
'{"Column": "temperatureAlert", "Properties": {"Path":
"$.Properties.temperatureAlert"}},' '{ "column" : "IotHubDeviceId",
"Properties":{"Path":"$.iothub-connection-device-id"}},'
'{"Column": "IotHubEnqueuedTime", "Properties": {"Path": "$.iothub-
enqueuedtime"}}' ']'
```

Now you are ready to add a data connection to IoT Hub collecting the messages. Clicking on **Data connections** in the left-side menu will bring up a page where you can add a data connection to IoT Hub. The following figure shows a connection that has already been created.

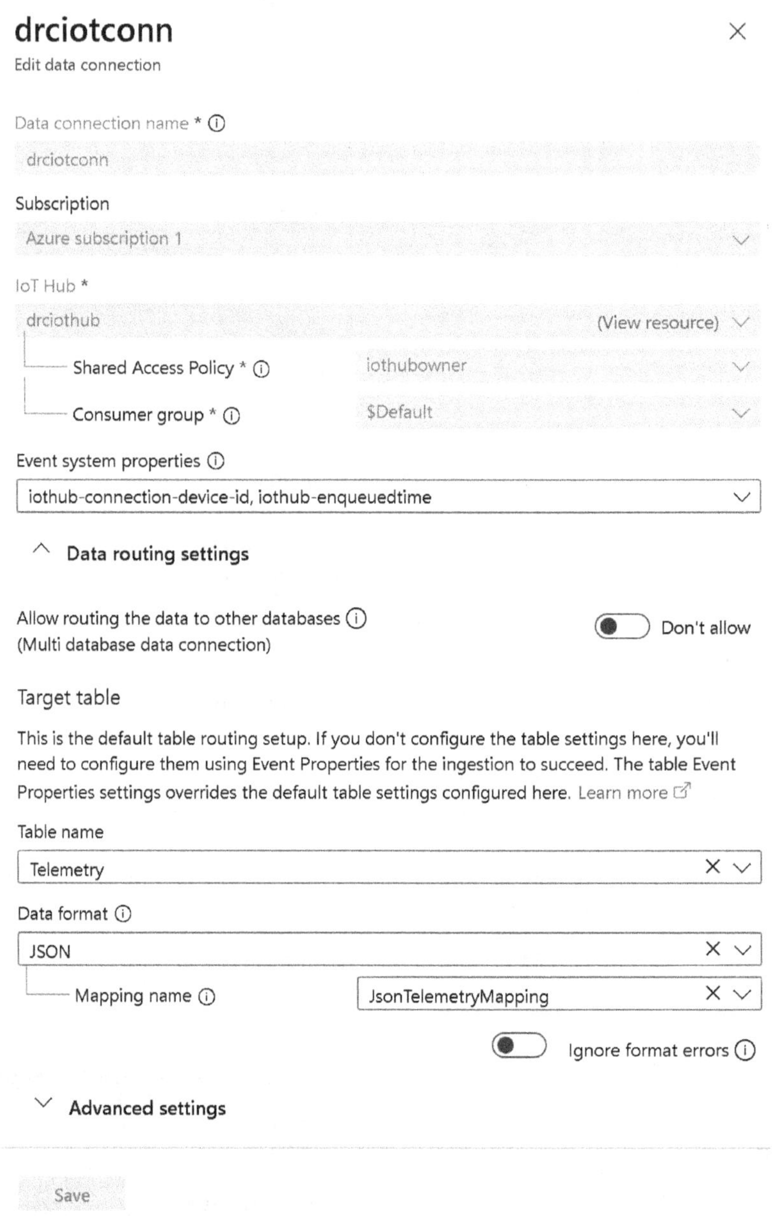

drciotconn

Edit data connection

Data connection name * ⓘ

drciotconn

Subscription

Azure subscription 1

IoT Hub *

drciothub   (View resource)

└──── Shared Access Policy * ⓘ   iothubowner

└──── Consumer group * ⓘ $Default

Event system properties ⓘ

iothub-connection-device-id, iothub-enqueuedtime

⌃ **Data routing settings**

Allow routing the data to other databases ⓘ   ⬤ Don't allow
(Multi database data connection)

Target table

This is the default table routing setup. If you don't configure the table settings here, you'll
need to configure them using Event Properties for the ingestion to succeed. The table Event
Properties settings overrides the default table settings configured here. Learn more ↗

Table name

Telemetry   ✕ ⌄

Data format ⓘ

JSON   ✕ ⌄

└──── Mapping name ⓘ   JsonTelemetryMapping   ✕ ⌄

                                                                ⬤ Ignore format errors ⓘ

⌄ **Advanced settings**

Save

Figure 8.3 – A data connection set up in ADX

To verify you are getting streaming data, switch to the query editor in ADX and issue the following KQL:

```
Telemetry
| count
```

After a few minutes, you can issue the query again and then the count should increase as the messages stream in. In the next section, we will look at visualizing the data.

Visualizing the streaming data

The visualization and reporting of data play a vital role in the data analytics workflow. ADX enables the development of advanced analytics solutions to handle extensive data volumes. By seamlessly integrating with multiple visualization tools, ADX allows you to effectively present and distribute data insights throughout your organization. These insights can be converted into actionable information to drive meaningful outcomes for your business.

Along with its own dashboarding capabilities, ADX integrates with other visualization tools such as Power BI, Excel, Grafana, Tableau, and Qlik, to name just a few. In addition, ADX provides native advanced analytics for time series analysis, pattern recognition, anomaly detection, and forecasting.

To create a dashboard in ADX, navigate to the ADX web UI and select the **Dashboard** tab in the left-side menu.

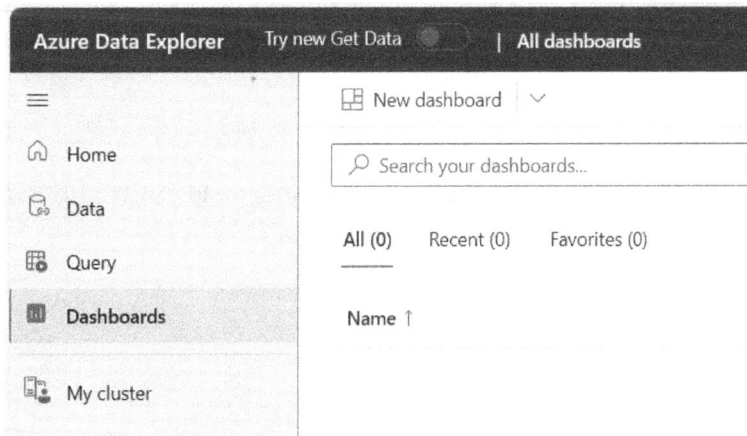

Figure 8.4 – Creating a dashboard

You get the option of creating a new dashboard or a sample dashboard. If you select the sample IoT dashboard, you will see the following dashboard:

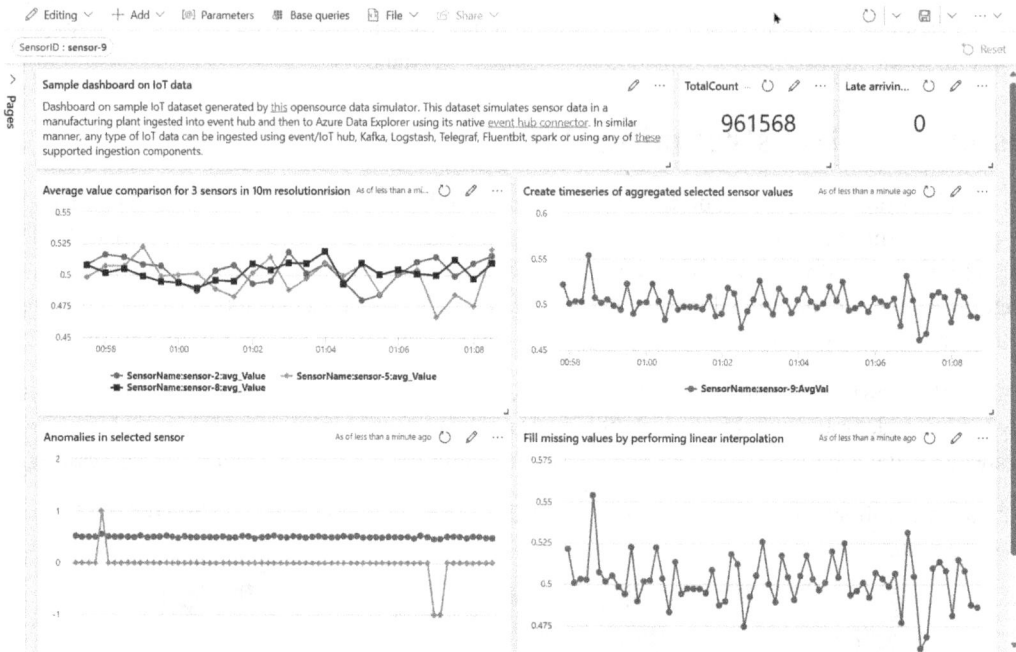

Figure 8.5 – The sample IoT dashboard

As you can see, ADX supports a variety of visualizations, useful for IoT analysis, and also includes the ability to create parameters useful for dynamic analysis.

Now that you have seen how IoT data can be visualized with ADX, it is time to get some hands-on experience with an exercise in the next section.

Lab – creating an ADX dashboard

In this lab, we will explore how to consume and analyze IoT data using ADX. We will then leverage ADX's capabilities to create a real-time dashboard that visualizes the IoT data:

1. Set up IoT Hub and a device capturing data from the Raspberry Pi online simulator (`https://azure-samples.github.io/raspberry-pi-web-simulator/#getstarted`).

2. Verify that the signal is being sent to IoT Hub.

3. Set up an SDX cluster and make sure you enable streaming data.

4. Create a database and a table in ADX to capture the data.

5. Define the table mapping between the incoming messages and the table columns.

6. Create a new dashboard and enter the following KQL using your table name:

```
Telemetry
 | where IotHubEnqueuedTime between (['_startTime'] .. ['_
endTime']) // Time range filtering
```

7. After a few minutes, you should see results when you run the KQL.

8. Select **Add visual** next to the **Results** tab.

9. Add a line chart showing the temperature.

10. Select **Apply changes**. Your final result should look similar to *Figure 8.6*.

11. When finished, delete your resources.

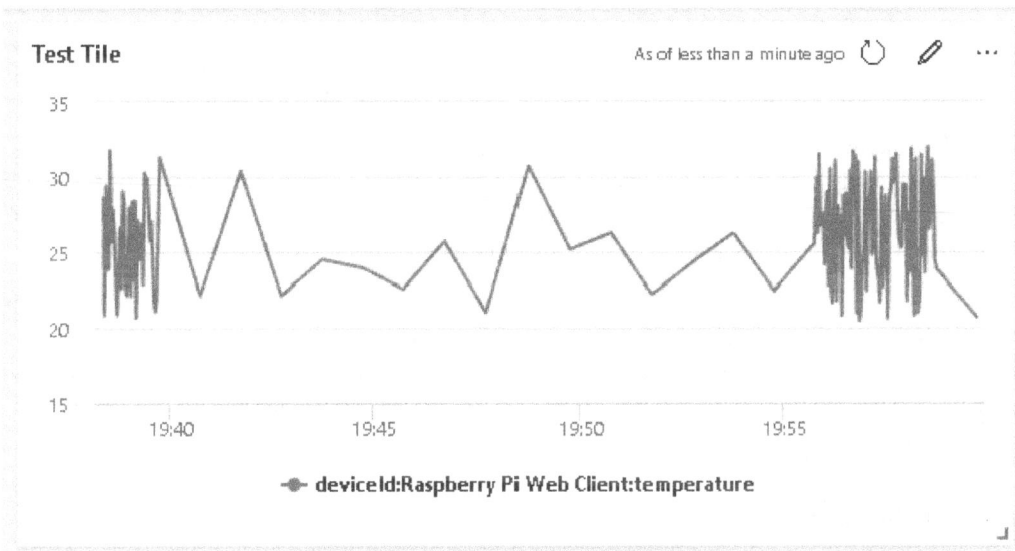

Figure 8.6 – Sample ADX dashboard tile

Now that you have created a tile in the ADX dashboard, you should be more comfortable working with ADX. We have just touched on part of what you can do in ADX (there are books just on this topic). I would highly recommend further study of ADX.

Summary

In this chapter, you were introduced to ADX, which is an excellent tool to use when exploring and analyzing the data collected from IoT devices. With the proliferation of IoT devices in various industries, there is an immense amount of data being generated, and the ability to extract valuable insights from this data is crucial.

The chapter began by introducing ADX and explaining its role in handling large volumes of streaming data. It highlighted ADX's capabilities in ingesting, storing, and querying data in real time, making it suitable for scenarios involving IoT devices, log analytics, and other streaming data sources. Next, the chapter dug into the process of ingesting streaming data into ADX. It provided guidance on configuring data ingestion pipelines.

Another crucial aspect of streaming data analysis is data visualization, and the chapter explored how ADX enables users to create compelling visual representations of their data. To provide you with hands-on experience, the chapter included a lab exercise on creating an ADX dashboard.

Overall, the chapter provided an overview of ADX's capabilities for streaming data analysis and visualization. It equipped you with the knowledge and practical skills needed to ingest, analyze, and visualize streaming data in real time using ADX. Whether it's monitoring IoT devices, analyzing log data, or conducting time series analysis, you should now have an understanding of how ADX can be a valuable tool in your data analytics toolkit.

So far, you have seen how to analyze data in the cloud, but there are times when it is beneficial to move some of the processing to your devices. Edge devices have extra compute and storage so you can install IoT edge modules on them. In the next chapter, you will learn when and how to set up and use edge devices in your IoT solution.

Further reading

This book is worth looking at if you wish to delve deeper into ADX and KQL:

- *Scalable Data Analytics with Azure Data Explorer, Packt Publishing*, by *Jason Myerscough*

9

Exploring IoT Edge Computing

In the ever-evolving landscape of technology, the **Internet of Things** (**IoT**) has emerged as a transformative force, connecting devices, sensors, and systems to enable seamless communication and data exchange. However, as the IoT continues to expand, it faces several challenges, including latency, bandwidth limitations, security concerns, and the sheer volume of data generated. To address these obstacles and unlock the true potential of the IoT, a revolutionary approach has emerged: **IoT Edge computing**.

In this chapter, we delve into the realm of IoT Edge computing, where intelligence and processing capabilities are pushed closer to the network edge. Rather than solely relying on cloud-based servers for data analysis and decision-making, IoT Edge computing brings computation, storage, and analytics closer to the devices and sensors themselves. By doing so, it empowers organizations and individuals to harness real-time insights, make timely decisions, and maximize the benefits of the IoT.

In this chapter, we will take a look at the following topics:

- Azure IoT Edge devices
- Deploying an IoT Edge device
- Exploring `EdgeAgent`
- Exploring `EdgeHub`
- Computing and storage on the edge
- Using an edge device as a gateway
- Lab – implementing stream analytics on the edge

Azure IoT Edge devices

Azure IoT Edge devices are physical devices that can run the Azure IoT Edge runtime, a component of the Azure IoT Edge service provided by Microsoft. Azure IoT Edge is a platform that enables users to deploy and manage cloud workloads on edge devices, bringing the power of the cloud closer to devices generating data.

Azure IoT Edge devices can include a wide range of devices, including industrial equipment, gateways, sensors, and cameras. These devices are typically located at the edge of a network, closer to where the data is generated, and have more compute and storage compared to typical IoT devices. By deploying the Azure IoT Edge runtime on these devices, users can run cloud services and AI models directly on the devices, enabling real-time data processing, analytics, and decision-making at the edge.

Some key features and capabilities of Azure IoT Edge devices include:

- **Edge computing**: Azure IoT Edge devices can process and analyze data locally, reducing latency and improving real-time decision-making capabilities

- **Secure communication**: Devices can securely communicate with the cloud through the Azure IoT Edge runtime, which provides authentication, encryption, and other security features

- **Module deployment**: Azure IoT Edge devices support the deployment of modules (units of code that can be containers or functions), enabling users to run custom logic, AI models, or third-party services on the edge device

- **Offline capabilities**: Azure IoT Edge devices can continue to operate even when disconnected from the cloud, allowing them to function in remote or intermittent connectivity scenarios

- **Device management**: Azure IoT Edge provides tools and services to remotely manage and monitor edge devices, including **over-the-air** (OTA) updates, configuration management, and diagnostics

By leveraging Azure IoT Edge devices, organizations can benefit from edge computing capabilities, enabling them to process and analyze data closer to the source, reduce data transfer costs, and make faster, intelligent decisions at the edge.

Now that you know some of the characteristics of IoT Edge devices, it is time to see how they get deployed.

Deploying an IoT Edge device

To deploy an Azure IoT Edge device, you typically need to follow these key steps:

1. **Set up an Azure IoT hub**: Create an Azure IoT hub, which acts as a central messaging hub for bi-directional communication between your edge devices and the cloud. This hub will manage device identities, security, and message routing, just as it does for the devices as well.

2. **Provision the IoT Edge device**: Register and provision your physical device as an Azure IoT Edge device. This involves creating a device identity within the IoT hub and generating the necessary security credentials (for example, device connection string or X.509 certificate) to authenticate the device with the hub. The following screenshot shows how to set up an IoT Edge device; you can see that you just have to check the box indicating the device is an IoT Edge device:

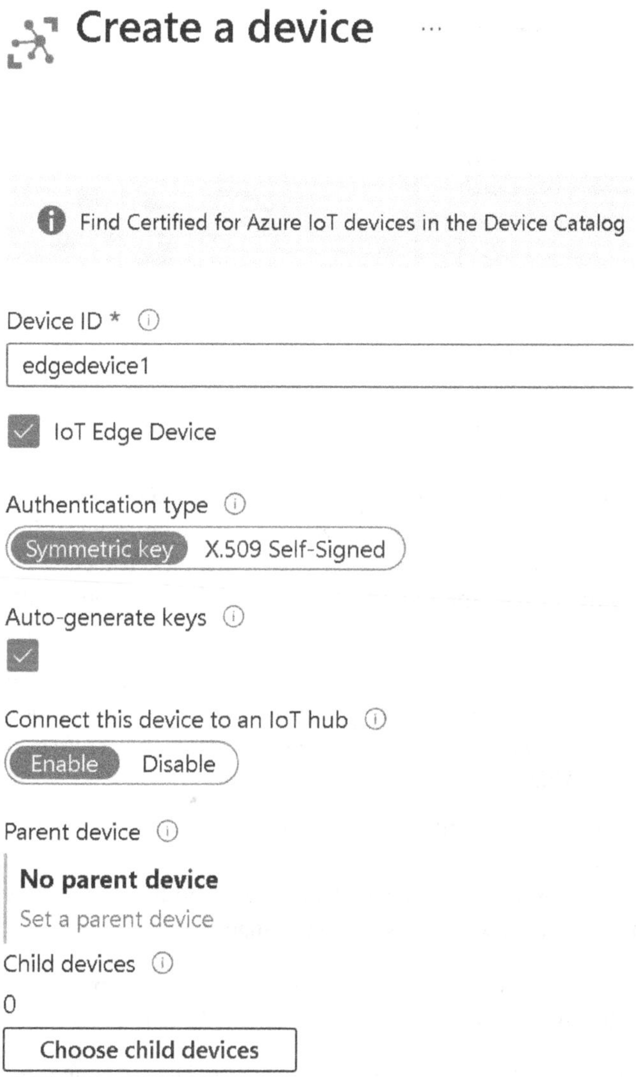

Create a device ...

Find Certified for Azure IoT devices in the Device Catalog

Device ID * ⓘ

edgedevice1

☑ IoT Edge Device

Authentication type ⓘ

(Symmetric key) X.509 Self-Signed

Auto-generate keys ⓘ

☑

Connect this device to an IoT hub ⓘ

(Enable) Disable

Parent device ⓘ

No parent device

Set a parent device

Child devices ⓘ

0

Choose child devices

Figure 9.1 – Provisioning an edge device

3. **Install and configure the Azure IoT Edge runtime:** Because you indicated this was an IoT Edge device, two modules need to be installed: the EdgeAgent and EdgeHub modules (we will delve into these modules in the upcoming subsections). Install the Azure IoT Edge runtime on your device. The runtime provides the execution environment for modules and manages communication between modules and the IoT hub. The following diagram shows the communication of the IoT runtime:

Azure IoT edge device

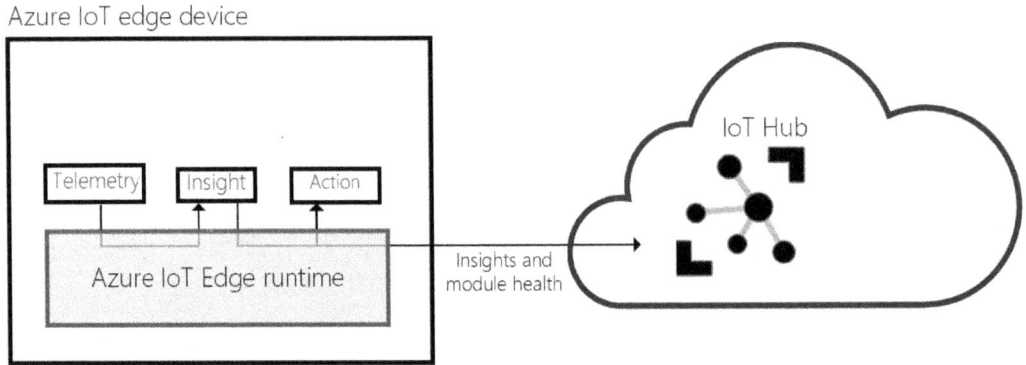

Figure 9.2 – The Azure IoT Edge runtime

4. **Create and deploy IoT Edge modules**: Develop or acquire modules that contain the code or AI models you want to run on the edge device. Modules can be created using various languages and frameworks, such as C#, Python, or Docker containers. You define the modules in the deployment manifest, which specifies the desired modules and their configurations.

5. **Deploy the IoT Edge deployment manifest**: Once you have created the deployment manifest that describes the modules and their configurations, you need to deploy it to the Azure IoT Edge device. The deployment manifest specifies the desired state of modules on the device. You can deploy the manifest through various means, such as the Azure portal, the Azure CLI, or Azure IoT Edge APIs.

6. **Monitor and manage the IoT Edge device**: Azure provides tools and services to monitor and manage your deployed Azure IoT Edge devices. You can use the Azure portal, the Azure CLI, or Azure IoT Hub APIs to monitor device health, view module metrics, perform diagnostics, and manage module updates or configuration changes remotely.

By following these steps, you can successfully deploy an Azure IoT Edge device and start leveraging the power of edge computing to process data and run workloads closer to the source of data generation.

As mentioned previously, the Azure IoT Edge agent is a crucial component in coordinating and managing IoT Edge components. In the next section, you will take a closer look at the `EdgeAgent` module and its function.

Exploring EdgeAgent

The EdgeAgent module is a crucial component of an Azure IoT Edge device and is part of the IoT runtime. Its purpose is to manage the life cycle and deployment of modules on the IoT Edge device. Here are some key functions and responsibilities of the EdgeAgent module:

- **Module management**: The EdgeAgent module is responsible for managing modules running on the IoT Edge device. It coordinates the deployment, starting, stopping, and removal of modules based on the desired state specified in the deployment manifest.

- **Communication with IoT Hub**: The EdgeAgent module establishes and maintains a secure connection with Azure IoT Hub. It acts as the intermediary between the IoT hub and the modules running on the edge device, facilitating bi-directional communication.

- **Deployment synchronization**: The EdgeAgent module continuously syncs with Azure IoT Hub to ensure that the deployment manifest is up to date. If there are changes in the desired state of modules (for example, updates, additions, or deletions), the EdgeAgent module manages the synchronization process and applies the changes to the device.

- **Health monitoring and reporting**: The EdgeAgent module monitors the health and status of modules running on the device. It periodically reports module status, metrics, and diagnostic information to Azure IoT Hub, enabling remote monitoring and management of the device.

- **Security and authentication**: The EdgeAgent module handles the security aspects of module-to-module and module-to-cloud communication. It manages authentication and access control for modules, ensuring that only authorized modules can communicate with each other or send data to the cloud.

- **Offline operation**: The EdgeAgent module enables offline operation of the IoT Edge device. It caches the deployment manifest and module configurations locally, allowing the device to continue functioning even when disconnected from the IoT hub. Once the device reconnects, the EdgeAgent module synchronizes any pending changes with the hub.

Overall, the EdgeAgent module plays a critical role in managing the modules and their life cycle on an IoT Edge device. It ensures that the desired state of modules is maintained, facilitates communication with the IoT hub, monitors module health, and provides security and authentication capabilities.

Exploring EdgeHub

The EdgeHub module is another core module within the Azure IoT Edge runtime, and its purpose is to enable communication and message routing between modules running on an IoT Edge device. Here are the key purposes and functionalities of the EdgeHub module:

- **Message routing**: The EdgeHub module acts as a message broker, facilitating communication between modules running on the IoT Edge device. It receives messages from the modules and routes them to the appropriate destination modules based on message routing rules defined in the deployment manifest.

- **Local communication**: The EdgeHub module allows modules to communicate with each other locally within a device. Modules can send messages to other modules running on the same device through EdgeHub, enabling modular and decoupled application architectures.

- **Cloud connectivity**: The EdgeHub module establishes a secure connection with Azure IoT Hub, enabling bi-directional communication between the IoT Edge device and the cloud. It forwards messages from modules to the cloud and vice versa, allowing data and commands to be exchanged between the device and cloud services.

- **Protocol translation**: The EdgeHub module supports various communication protocols, such as MQTT, AMQP, and HTTP. It can perform protocol translation, allowing modules that use different protocols to seamlessly communicate with each other or with cloud services.

- **Offline support**: The EdgeHub module provides offline capabilities for the IoT Edge device. It stores messages locally when the device is disconnected from the cloud and forwards them when the connection is restored, ensuring message delivery even in intermittent or low-connectivity scenarios.

- **Security and authentication**: The EdgeHub module handles the security aspects of communication between modules and the cloud. It supports secure and encrypted communication channels, and it manages the authentication and access control for modules accessing the EdgeHub module.

- **Message persistence and buffering**: The EdgeHub module can buffer and store messages in case of network disruptions or module failures. This ensures that messages are not lost and can be reliably delivered to the intended modules or forwarded to the cloud when connectivity is restored.

In summary, the EdgeHub module provides message routing, local communication, cloud connectivity, protocol translation, offline support, and security capabilities within the Azure IoT Edge runtime. It plays a vital role in facilitating communication and data exchange between modules running on an IoT Edge device and enables seamless integration with cloud services.

Now that we have discussed the required modules we need to facilitate edge processing, let us look at using a compute and a storage module on the edge.

Computing and storage on the edge

There are many different modules you can download and install on your IoT Edge devices. These include modules supplied by Microsoft and third-party industry-specific modules. You can even create your own module to use on your devices. These include modules for blob storage, SQL Database, video analysis, and analytics, to name just a few.

Installing these in the Azure portal is pretty straightforward. Navigate to the IoT hub device and then go to **Set modules on device**. From there, you can select an IoT Edge module to install:

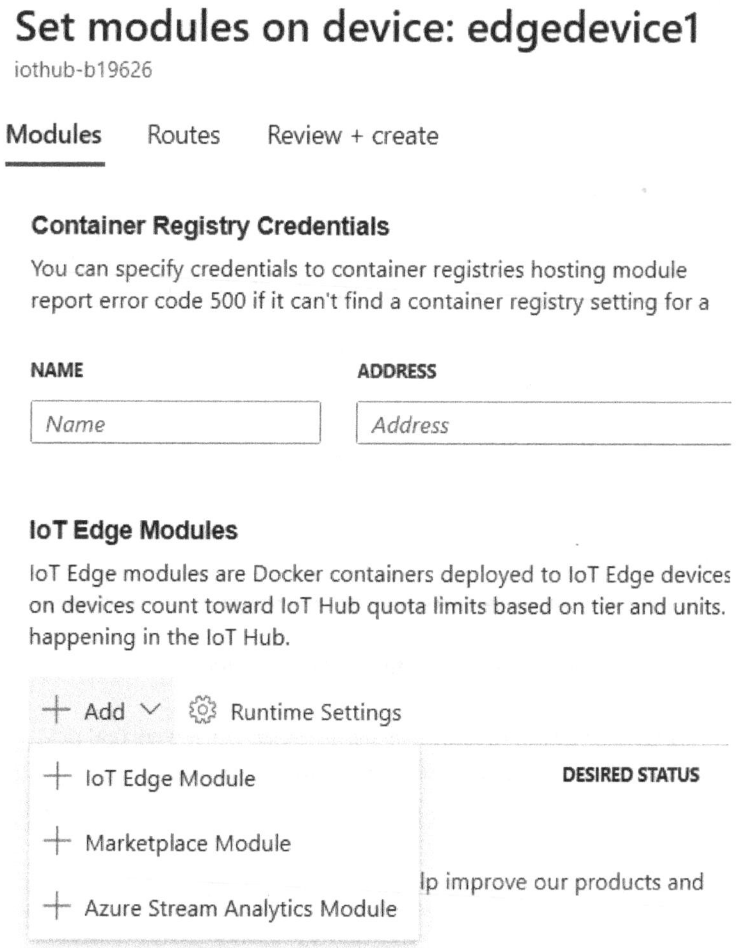

Set modules on device: edgedevice1
iothub-b19626

Modules   Routes   Review + create

Container Registry Credentials
You can specify credentials to container registries hosting module
report error code 500 if it can't find a container registry setting for a

NAME

Name

ADDRESS

Address

IoT Edge Modules
IoT Edge modules are Docker containers deployed to IoT Edge devices
on devices count toward IoT Hub quota limits based on tier and units.
happening in the IoT Hub.

+ Add ∨   ⚙ Runtime Settings

+ IoT Edge Module   **DESIRED STATUS**

+ Marketplace Module
                                 lp improve our products and
+ Azure Stream Analytics Module

Figure 9.3 – Setting up modules on an edge device

The easiest option is to select **Marketplace Module**. This will show you modules developed by Microsoft and various third-party software developers:

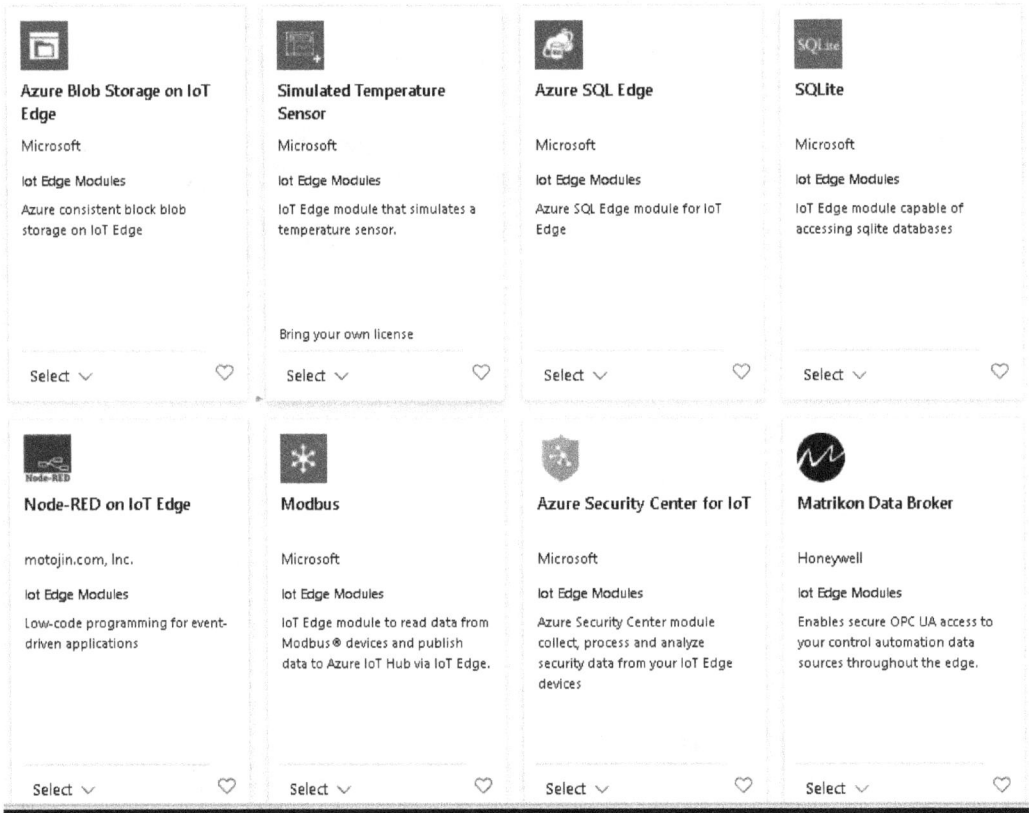

Azure Blob Storage on IoT Edge

Microsoft

Iot Edge Modules

Azure consistent block blob storage on IoT Edge

Select ⌄

Simulated Temperature Sensor

Microsoft

Iot Edge Modules

IoT Edge module that simulates a temperature sensor.

Bring your own license

Select ⌄

Azure SQL Edge

Microsoft

Iot Edge Modules

Azure SQL Edge module for IoT Edge

Select ⌄

SQLite

Microsoft

Iot Edge Modules

IoT Edge module capable of accessing sqlite databases

Select ⌄

Node-RED on IoT Edge

motojin.com, Inc.

Iot Edge Modules

Low-code programming for event-driven applications

Select ⌄

Modbus

Microsoft

Iot Edge Modules

IoT Edge module to read data from Modbus® devices and publish data to Azure IoT Hub via IoT Edge.

Select ⌄

Azure Security Center for IoT

Microsoft

Iot Edge Modules

Azure Security Center module collect, process and analyze security data from your IoT Edge devices

Select ⌄

Matrikon Data Broker

Honeywell

Iot Edge Modules

Enables secure OPC UA access to your control automation data sources throughout the edge.

Select ⌄

Figure 9.4 – Viewing the Marketplace module

In this chapter, we will use a **virtual machine** (**VM**) to act as our IoT virtual device. The VM has the IoT Edge runtime installed on it. After installing a module, you can check its status on the edge device in the portal:

Modules   IoT Edge hub connections   Deployments and Configurations

Name	Type	Specified in Deployment	Reported by Device	Runtime Status	Exit Code
$edgeAgent	IoT Edge System Module	✓ Yes	✓ Yes	running	NA
$edgeHub	IoT Edge System Module	✓ Yes	✓ Yes	running	NA
AzureBlobStorageonIoTEdge	IoT Edge Custom Module	✓ Yes	✓ Yes	running	NA

Figure 9.5 – Checking the status of the edge runtime

You can install more than one module. For example, if you install a Blob Storage module, you can create another module (maybe a Stream Analytics module) to process the signals and place them in the blob storage module. The following .NET classes are used to interact with the blob storage:

- `BlobServiceClient`: The `BlobServiceClient` class lets you manipulate Azure Storage resources and blob containers

- `BlobContainerClient`: The `BlobContainerClient` class permits you to manipulate Azure Storage containers and their blobs

- `BlobClient`: The `BlobClient` class allows you to manipulate Azure Storage blobs

Now that you have seen how to create an edge device and use modules on the edge device, there is another common use case for edge devices you should investigate. That is setting up gateway devices used for protocol translation, which we will look at next.

Using an edge device as a gateway

An IoT Edge device performs protocol translation by acting as an intermediary that can understand and convert data between different communication protocols used by IoT devices and the central data center or cloud. This capability is crucial in IoT systems where various devices may communicate using different protocols, and it's necessary to establish interoperability and ensure a seamless data flow between them.

Here's how an IoT Edge device typically performs protocol translation:

1. **Data collection**: The IoT Edge device first collects data from the local IoT devices within its network. These devices could be using different communication protocols, such as MQTT, CoAP, Zigbee, Z-Wave, Modbus, Bluetooth, LoRaWAN, or various proprietary protocols.

2. **Protocol detection**: Upon receiving data from the IoT devices, the edge device examines the incoming data packets to identify the source communication protocol. It needs to recognize which protocol the data is currently using to understand how to handle and translate it properly.

3. **Protocol translation**: Once the edge device identifies the source protocol, it uses the appropriate translation mechanism to convert the data into a standardized format or another protocol that is understood by the central data center or cloud. This translation could involve changing data structures, headers, addressing schemes, or other protocol-specific details.

4. **Data preprocessing** (*optional*): In some cases, the edge device may perform additional data preprocessing or aggregation before translating the data. This step can involve filtering out irrelevant information or performing basic data analytics to reduce the amount of data that needs to be sent to the cloud, optimizing bandwidth usage and reducing latency.

5. **Data transmission**: After the protocol translation and optional preprocessing, the edge device sends the converted data to the cloud or central data center using the appropriate communication protocol that the cloud can understand, typically using protocols such as HTTP/HTTPS, MQTT, AMQP, or others commonly used in cloud environments.

6. **Reverse translation** (*optional*): If the cloud sends commands or responses back to the IoT devices, the edge device can also perform reverse translation to ensure that the instructions are in the format compatible with the respective IoT devices' protocols.

It's essential for IoT Edge devices to support multiple communication protocols to handle the diverse range of IoT devices they might encounter in real-world deployments. Additionally, as the IoT landscape evolves, edge devices may need to be updated to support new protocols or versions of existing protocols to maintain compatibility with the latest IoT devices and standards.

By performing protocol translation, IoT Edge devices enable seamless communication and data exchange between heterogeneous IoT devices and the central cloud, facilitating the integration and scalability of IoT solutions:

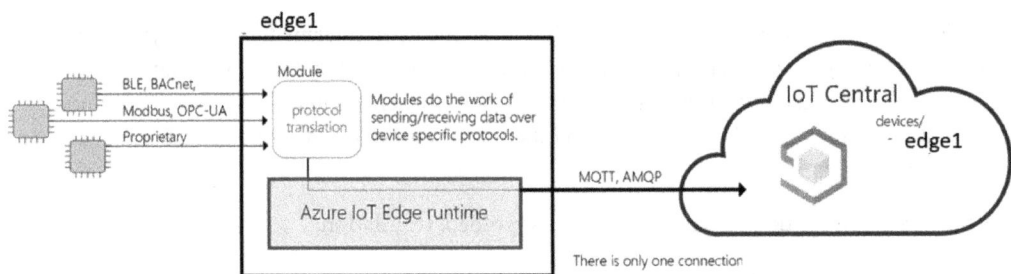

Figure 9.6 – Protocol translation

Now that you've reviewed the theory behind setting up and using modules on the edge, it is time to gain some hands-on experience in a lab.

Lab – implementing stream analytics on the edge

In this lab, you will do the following:

- Create an **Azure Stream Analytics (ASA)** job configured to process data on the edge

- Connect the ASA job with other IoT Edge modules running on the same edge device

- Configure the data flow between the Stream Analytics module and other modules using IoT Edge routes

Let's accomplish the aforementioned tasks by following these steps:

1. Go to your Azure subscription, open up the Cloud Shell, and make sure you are in the Bash environment. If this is your first time opening up the Cloud Shell in the portal, you will be asked to create an Azure Storage account to support the Cloud Shell.

2. Run the following command in the Cloud Shell to create a resource group:

    ```
    az group create –name IoTEdgeResources --location eastus2
    ```

    You can change the location to a supported one closer to you (for a list of region capabilities, see here: https://azure.microsoft.com/en-us/explore/global-infrastructure/geographies/#overview).

3. Create an IoT hub using the following command. Make sure you replace the hub name with your own value:

    ```
    az iot hub create --resource-group IoTEdgeResources --name {hub_
    name} --sku F1 --partition-count 2
    ```

4. Enter the following command to create and register an edge hub device:

    ```
    az iot hub device-identity create --device-id myEdgeDevice
    --edge-enabled --hub-name {hub_name}
    ```

5. Enter the following command to create a VM that will simulate our device. Make sure you replace the values with your values where indicated:

    ```
    az deployment group create \
    --resource-group IoTEdgeResources \
    --template-uri "https://raw.githubusercontent.com/Azure/iotedge-
    vm-deploy/1.4/edgeDeploy.json" \
    --parameters dnsLabelPrefix='<REPLACE_WITH_VM_NAME>' \
    --parameters adminUsername='azureUser' \
    --parameters deviceConnectionString=$(az iot hub device-identity
    connection-string show --device-id myEdgeDevice --hub-name
    <REPLACE_WITH_HUB_NAME> -o tsv) \
    --parameters authenticationType='password' \
    --parameters adminPasswordOrKey="<REPLACE_WITH_PASSWORD>"
    ```

6. To add modules to your device, in the Azure portal, open your IoT hub and select **Devices** on the left-hand menu.

7. Select the name of your device from the device list.

8. On the device page, you may see a warning stating the `status code 417 -- The device's deployment configuration is not set` in the Azure portal. Don't worry, as this is normal when a device is first created.

9. Go to the **Set modules** tab at the top of the page. Add a Marketplace module by selecting it in the dropdown.

10. Search the marketplace for the `Simulated Temperature Sensor` module and add it to the device.

11. After adding the module, you will be returned to the device details page.

12. You should see three modules: `$edgeAgent`, `$edgeHub`, and `SimulatedTemperatureSensor`. Wait until they are up and running on your device (this may take a few minutes).

13. Create a Stream Analytics job in your resource group. Be sure to designate it as an edge-hosted service.

14. Once up and running, add an edge hub as the input.

15. For the output, add another edge hub.

16. Add the following code for the query. This query issues a reset message whenever the temperature exceeds 70 degrees in the previous 30 seconds. Replace the input and output with your values:

```
SELECT  'reset' AS command
 [your output]
 FROM
 [your input TIMESTAMP BY timeCreate
 GROUP BY TumblingWindow(second,30)
 HAVING Avg(machine.temperature) > 70
```

17. Go back to **Set modules** on the IoT Edge device screen in the portal. This time, select a Stream Analytics job. Point it to the Stream Analytics job you just created.

18. Go to the **Set modules on device: <your-device-name>** page, and click on **Next: Routes**.

    On the **Routes** tab, you will specify how messages are transmitted between modules and the IoT hub. Messages will be constructed using name and value pairs.

19. Set up the routes by using the names and values indicated in the following table. Replace `{moduleName}` with the actual name of your Azure Stream Analytics module. Ensure that the module name matches the one listed in the modules section of your device on the **Set modules** page in the Azure portal:

```
Route Name    | Value
------------- | ----------------------
telemetryToCloud | FROM /messages/modules/
SimulatedTemperatureSensor/* INTO $upstream
alertsToCloud | FROM /messages/modules/{moduleName}/* INTO
$upstream
alertsToReset | FROM /messages/modules/{moduleName}/* INTO
BrokeredEndpoint("/modules/SimulatedTemperatureSensor/inputs/
control")
```

```
telemetryToAsa | FROM /messages/modules/
SimulatedTemperatureSensor/* INTO BrokeredEndpoint("/modules/
{moduleName}/inputs/temperature")
```

20. Proceed to the **Next: Review + Create** step. In the **Review + Create** tab, you can see how the information you provided in the wizard is translated into a JSON deployment manifest.

21. Once you have reviewed the manifest, click on **Create** to complete the configuration of your module.

    On the **Set modules** page of your device, after a few minutes, you should see the modules listed and running. Refresh the page if you don't see modules, or wait a few more minutes, then refresh it again.

22. To view the messages being sent to the IoT hub, go to the **Direct Method** tab and issue the UploadModuleLogs method with the following JSON payload (which just defines how the message will be sent). Remember to replace <sasUrl> with the URL valid for your device:

```
{
    "schemaVersion": "1.0",
    "sasUrl": "<sasUrl>",
    "items": [
        {
            "id": "edgeAgent",
            "filter": {
                "tail": 10
            }
        }
    ],
    "encoding": "none",
    "contentType": "text"
}
```

23. On the **Set modules** page of your device, after a few minutes, you should see the modules listed and running. Refresh the page if you don't see modules, or wait a few more minutes, then refresh it again.

24. When finished, delete your resources.

And with that, we have arrived at the end of this chapter.

Summary

In this chapter, we explored the concept of IoT Edge computing, which involves moving intelligence and processing capabilities closer to the network edge. Unlike traditional reliance on cloud-based servers for data analysis and decision-making, IoT Edge computing enables computation, storage, and analytics to be brought nearer to the devices and sensors. This approach empowers both organizations and individuals to leverage real-time insights, make prompt decisions, and optimize the advantages of the IoT.

In today's data-driven landscape, the ability to gain insights from real-time data is more critical than ever. As organizations strive to make data-informed decisions and respond swiftly to changing conditions, the demand for effective tools to visualize streaming data has surged. The next chapter delves into the exciting world of visualizing streaming data using Power BI, a powerful **business intelligence** (**BI**) and data visualization platform developed by Microsoft.

Streaming data, characterized by its continuous and rapid arrival, presents unique challenges and opportunities for analysis. Whether you are monitoring sensor data from IoT devices, tracking social media trends, or analyzing financial market fluctuations, harnessing the potential of real-time data can offer a competitive edge. The next chapter will equip you with the knowledge and skills to leverage Power BI's streaming capabilities, transforming your ability to make informed decisions and respond to changes as they happen.

Part 3:
Processing the Data

In the third part of the book, our focus turns to presenting data to users, using a tool such as Power BI to build dashboards, showing streaming data in real-time as it is captured. We also look at how we can incorporate machine learning and AI services to augment our insight into raw data streams. We end this part with a look at responding to device events, using Azure Data Grid. This service, along with an IoT hub, allows you to create a more responsive, scalable, and efficient IoT solution, simplifying event handling and enhancing security.

This part has the following chapters:

- *Chapter 10, Visualizing Streaming Data in Power BI*
- *Chapter 11, Integrating Machine Learning*
- *Chapter 12, Responding to Device Events*

10

Visualizing Streaming Data in Power BI

In the ever-evolving landscape of data-driven decision-making, the ability to access and visualize data in real time has become an indispensable asset for organizations across the globe. In this chapter, we will delve into the fundamental aspects of building Power BI dashboards that provide a window into the pulse of your data at any given moment.

Our journey begins with a solid grasp of Power BI dashboard basics, exploring the core elements that make up effective dashboards. Then, we'll dive into the world of real-time datasets, exploring the types of real-time datasets available and the unique advantages they offer. Whether you're dealing with streaming data, IoT devices, or APIs, understanding the various data sources at your disposal is essential to harnessing real-time insights effectively.

With a solid foundation in place, we'll turn our attention to pushing data into datasets. This section will guide you through the process of seamlessly integrating your data sources with Power BI, ensuring that your dashboards stay up to date and provide the most current information to your users.

To put theory into practice, we've included a hands-on lab – *Creating a real-time streaming dashboard*. In this lab, you'll have the opportunity to apply the concepts discussed throughout the chapter and build your own real-time streaming dashboard from scratch. By the end of this chapter, you'll be well-equipped to design, develop, and deploy real-time Power BI dashboards that empower your organization to make data-driven decisions with confidence. In particular, we will cover the following:

- Introducing Power BI
- Exploring Power BI dashboard basics
- Looking at the types of real-time datasets
- Pushing data into datasets
- Displaying real-time visuals
- Lab – creating a real-time streaming dashboard

Technical requirements

To follow along in this chapter, you will need either a Power BI license or a trial license. You can sign up for a license at `https://learn.microsoft.com/en-us/power-bi/fundamentals/service-self-service-signup-for-power-bi`.

Introducing Power BI?

Power BI is a business intelligence and data visualization tool developed by Microsoft. It allows users to connect to various data sources, transform and clean data, create interactive reports and dashboards, and share these insights with others. Power BI is widely used in organizations to analyze data, make data-driven decisions, and communicate information effectively.

Here are some key features and components of Power BI:

- **Data connectivity**: Power BI can connect to a wide range of data sources, including databases, cloud services, spreadsheets, and web services. It supports both on-premises and cloud-based data sources.

- **Data transformation**: Users can shape and transform data within Power BI using the Power Query Editor, enabling data cleansing, aggregation, and transformation operations.

- **Data modeling**: Power BI allows you to create data models by defining relationships between different tables, and creating calculated columns and measures. This helps in creating meaningful and interactive reports.

- **Report authoring**: Users can design interactive reports and dashboards using a drag and drop interface. You can add charts, tables, maps, and other visuals to represent data effectively.

- **Data Analysis Expressions (DAX)**: Power BI uses a formula language called DAX to create custom calculations and aggregations in your data model. It's similar to Excel formulas and allows complex calculations.

- **Data sharing**: Reports and dashboards created in Power BI can be shared with others within your organization or externally. You can publish reports to the Power BI service or export them as PDFs or PowerPoint presentations.

- **Collaboration**: Power BI supports collaboration features such as comments, notifications, and shared workspaces, making it easier for teams to work together on data analysis and reporting projects.

- **Security**: Power BI provides robust security and data protection features, allowing administrators to control access to data and reports based on roles and permissions.

- **Integration**: It integrates seamlessly with other Microsoft products such as Excel, SharePoint, and Azure services. It also supports third-party integrations and custom development through APIs.

- **Mobile accessibility**: Power BI provides mobile applications for a range of platforms, allowing users to conveniently access and engage with reports and dashboards on their mobile devices, such as smartphones and tablets.

Power BI comes in several versions, including **Power BI Desktop** (*for report creation*), the **Power BI service** (*for cloud-based sharing and collaboration*), and **Power BI Mobile** (*for accessing reports on mobile devices*). It is used across industries for various purposes, such as data analysis, business performance monitoring, and data-driven decision-making.

As you can see, there is a lot to Power BI, and a whole book could be written on this tool. In this chapter, we will concentrate on using the Power BI service to create a real-time dashboard to visualize streaming data.

Exploring Power BI dashboard basics

A **Power BI dashboard** is a visual and interactive representation of data and information created using Microsoft's Power BI platform. It provides a consolidated view of key metrics, reports, and data visualizations in a single, customizable interface. Power BI dashboards are designed to help users monitor and analyze data trends, make informed decisions, and gain insights into their business or organization.

Retail Analysis Sample

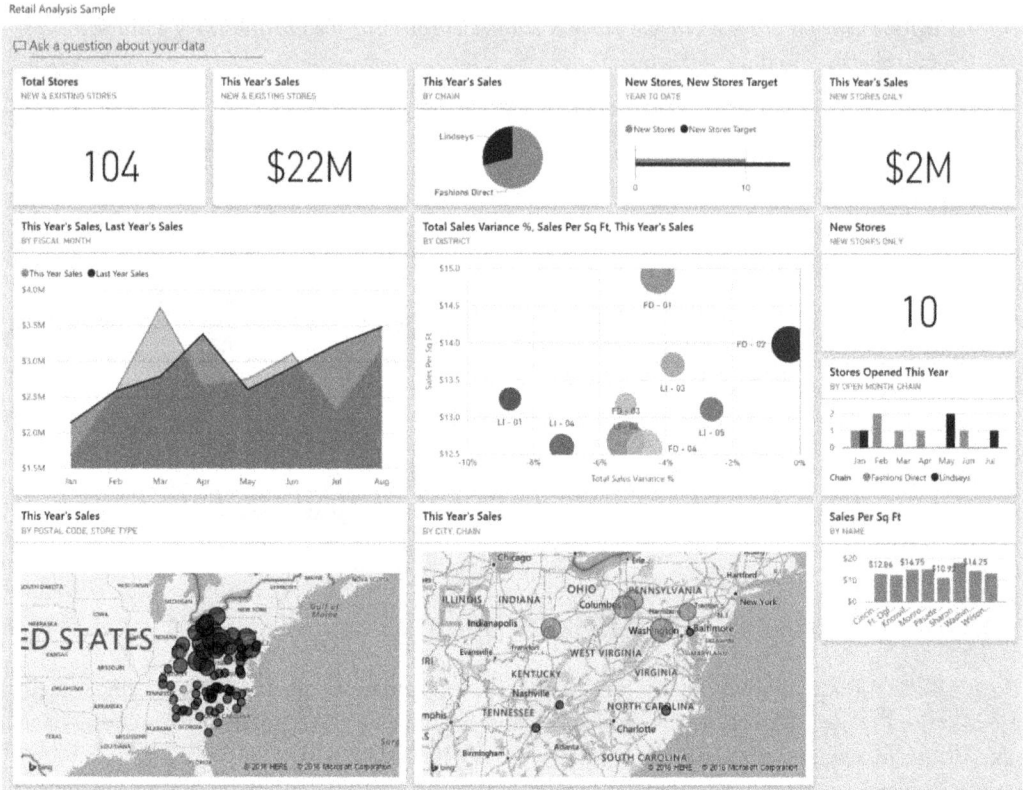

Figure 10.1 – A sample retail sales dashboard

Dashboards are composed of individual tiles, which are containers for specific data visualizations, such as charts, tables, maps, and images. Each tile displays a specific aspect of your data.

Power BI dashboards can be configured to display real-time or near-real-time data, making them suitable for monitoring live data streams and dynamic business situations.

Power BI allows users to set up data alerts, which trigger notifications when specified conditions or thresholds are met. This feature helps users stay informed of critical changes in their data.

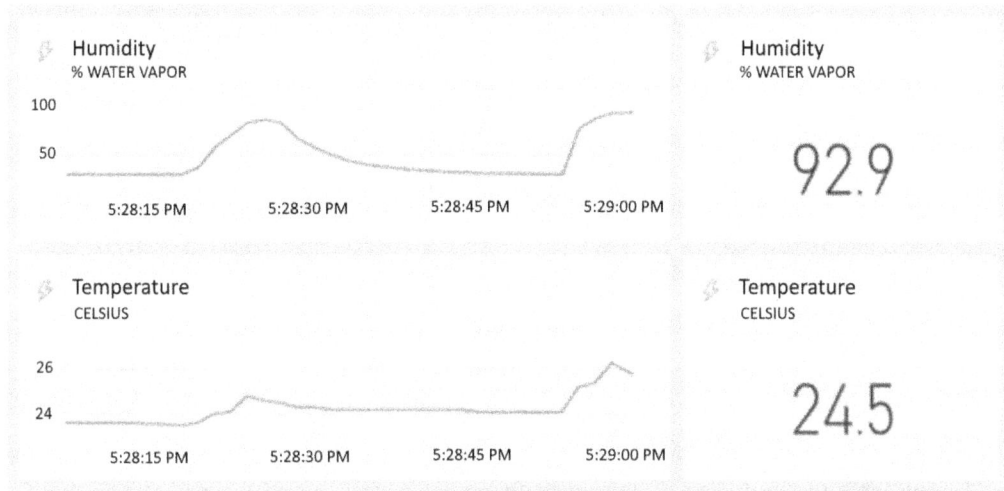

Figure 10.2 – A real-time dashboard showing IoT sensor data

Power BI dashboards are particularly useful for executives, managers, and decision-makers who need a concise and visually appealing way to monitor **key performance indicators** (**KPIs**) and track the health of their business processes. These dashboards provide a centralized location for data-driven insights, helping organizations make timely and informed decisions to drive their success.

To effectively harness the potential of real-time visuals, let's delve into a discussion of the various types of real-time datasets at your disposal.

Types of real-time datasets

Real-time datasets are collections of data that are continuously updated or streamed in real time, often with minimal or no delay. These datasets are valuable for various applications, including monitoring, analysis, and decision-making, in fields such as finance, healthcare, and transportation. Here are some common types of real-time datasets:

- **Financial market data**: Real-time financial datasets provide up-to-the-second information about stock prices, currency exchange rates, commodities, and market indices. They are essential for traders, analysts, and investors to make timely investment decisions.

- **IoT (Internet of Things) Data**: IoT devices generate vast amounts of real-time data, including sensor readings, temperature measurements, and GPS coordinates. These datasets are used in industries like smart cities, manufacturing, agriculture, and healthcare for monitoring and control.

- **Social media feeds**: Real-time social media datasets include posts, tweets, comments, and likes on platforms such as Twitter, Facebook, and Instagram. They are valuable for sentiment analysis, brand monitoring, and understanding public opinion in real time.

- **Weather data**: Real-time weather datasets provide continuous updates on weather conditions, including temperature, humidity, wind speed, and precipitation. Meteorologists, emergency services, and agriculture rely on this data for forecasting and decision-making.

- **Traffic and transportation data**: Real-time traffic data includes information about vehicle speed, congestion, accidents, and road closures. It is used to optimize transportation systems and GPS navigation and manage traffic flow.

- **Healthcare monitoring**: Real-time healthcare datasets can include patient vital signs, telemetry data, and electronic health records. These datasets help healthcare professionals monitor patients' conditions and respond quickly to critical events.

- **Environmental data**: Real-time environmental datasets track pollution levels, air quality, water quality, and other environmental parameters. They are important for environmental monitoring, regulatory compliance, and public health.

- **E-commerce and online retail**: Real-time e-commerce datasets capture customer behavior, transaction data, and website interactions. Online retailers use this data for personalization, inventory management, and fraud detection.

- **Energy usage data**: Real-time energy datasets monitor electricity, gas, and water consumption in homes and businesses. They are used for energy management and conservation efforts.

- **Network and server logs**: IT professionals use real-time network and server logs to monitor system performance, detect security breaches, and troubleshoot issues in real time.

- **Streaming media and video**: Real-time streaming datasets include live video feeds, online gaming data, and streaming media content. These datasets require low latency for a smooth user experience.

- **Supply chain and logistics**: Real-time supply chain data includes information on inventory levels, shipment tracking, and order processing. It helps companies optimize logistics operations and respond to supply chain disruptions.

- **Emergency services data**: Emergency services, such as police, fire, and medical responders, use real-time data for incident management, dispatching, and coordination during emergencies.

These are just a few examples of the types of real-time datasets that exist. We are concentrating on IoT, but as you can see, the specific datasets and their applications can vary widely across different industries and domains, but the common thread is the need for timely and continuous data updates to support decision-making and monitoring in real-time or near-real-time environments.

Now, let's see how we can integrate the data into Power BI.

Pushing data into datasets

The **push dataset** function operates by transferring data directly into the Power BI service. Upon creating the dataset, the Power BI service automatically establishes a new database within the service to house this incoming data.

Since there exists an underlying database that continuously stores the arriving data, you can construct reports using this data. These reports and their associated visualizations function similarly to other report visuals within Power BI. You can leverage all of Power BI's report-building capabilities, including Power BI visuals, data alerts, and pinning dashboard tiles.

Once you've crafted a report using the push dataset, you have the option to pin any of the report visuals onto a dashboard. These visuals on the dashboard will update in real time as data receives updates. Within the Power BI service, the dashboard initiates a tile refresh each time new data arrives.

The **streaming dataset** feature shares a similarity with push datasets in that it transfers data into the Power BI service. However, a crucial distinction exists – *Power BI temporarily caches the data in the case of streaming datasets, and this cache has a short lifespan*. This temporary cache serves the purpose of displaying visuals that have a fleeting historical context, such as a line chart covering a one-hour time window.

Unlike push datasets, a streaming dataset lacks an underlying database, preventing the creation of report visuals using the streaming data flow. Consequently, you cannot utilize report functionalities such as filtering, Power BI visuals, and other report-related features.

The sole method for visualizing a streaming dataset involves adding a tile and employing the streaming dataset as a custom streaming data source. Custom streaming tiles linked to a streaming dataset are designed to swiftly present real-time data. There is minimal delay between the data being pushed into the Power BI service and the visual update, as there's no requirement to write or read data from a database.

In practical terms, it is advisable to employ streaming datasets and their associated streaming visuals in scenarios where minimizing the time between data push and visualization is of utmost importance. Data should be pushed in a format that can be directly visualized without further aggregations. Examples of such readily visualizable data include temperature readings and pre-calculated averages.

PubNub is a real-time data streaming and messaging platform that enables developers to build applications with real-time capabilities, such as chat applications, live dashboards, and multiplayer games. PubNub provides a cloud-based infrastructure for real-time communication, making it easier to implement features such as real-time data synchronization and messaging in applications.

> **A PubNub streaming dataset**
>
> A **PubNub streaming dataset** typically refers to a dataset or data source that utilizes PubNub's real-time data streaming capabilities, continuously sending data from a source to multiple consumers or subscribers in real time. This data can include various types of information, such as sensor data, chat messages, location updates, or any other data that needs to be transmitted and updated in real time.

Now, let's take a look at creating and displaying real-time visuals on a dashboard.

Displaying real-time visuals

Follow these steps to create and display real-time visuals on a dashboard:

1. The first step in creating a dashboard is logging into the Power BI service (`https://app.powerbi.com/`).

2. Once in the service, you need to create a workspace. If you are asked to try the Power BI Pro trial, select **Yes**. This will give you 60 days of Pro features at no extra cost. Once created, you will get a blank workspace.

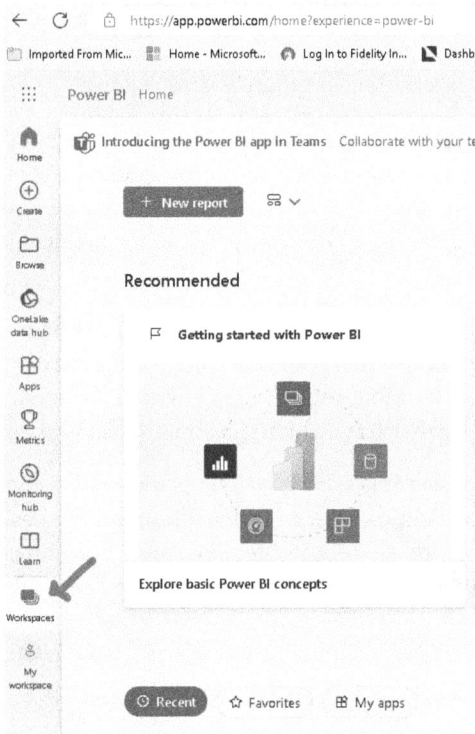

Figure 10.3 – Creating a workspace

3. To create a new dashboard, click on **New | Dashboard**.

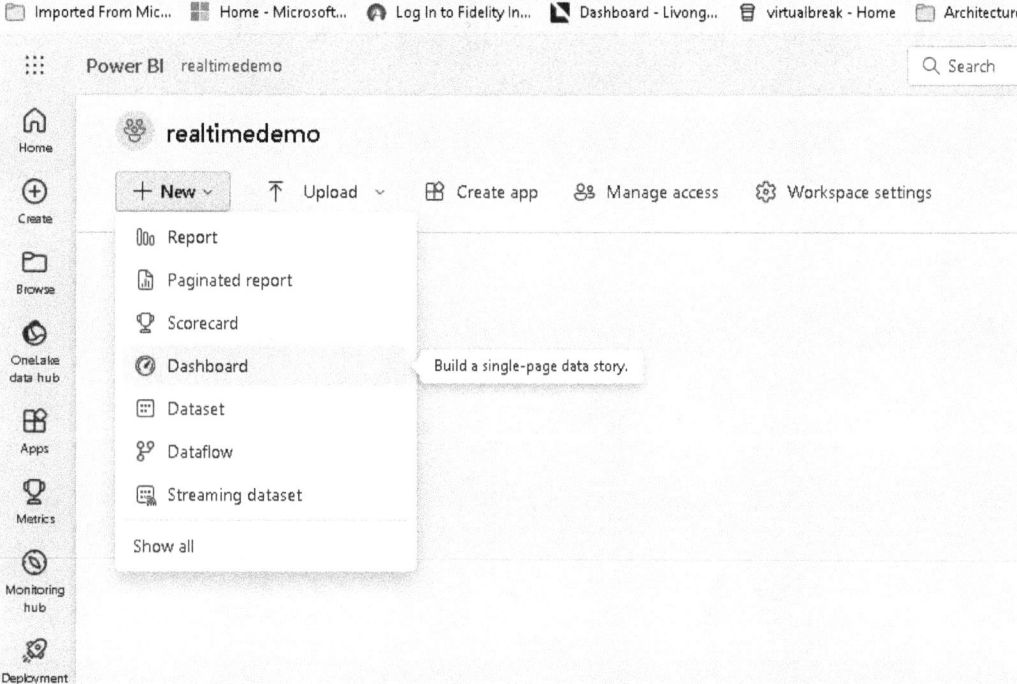

Figure 10.4 – Creating a dashboard

4.   Under the **Edit** dropdown, we can add tiles. A dashboard is made up of tiles that show different information. Along with pinning visuals from various reports, you can pin tiles that contain web content, images, and videos. At the bottom is the streaming data file.

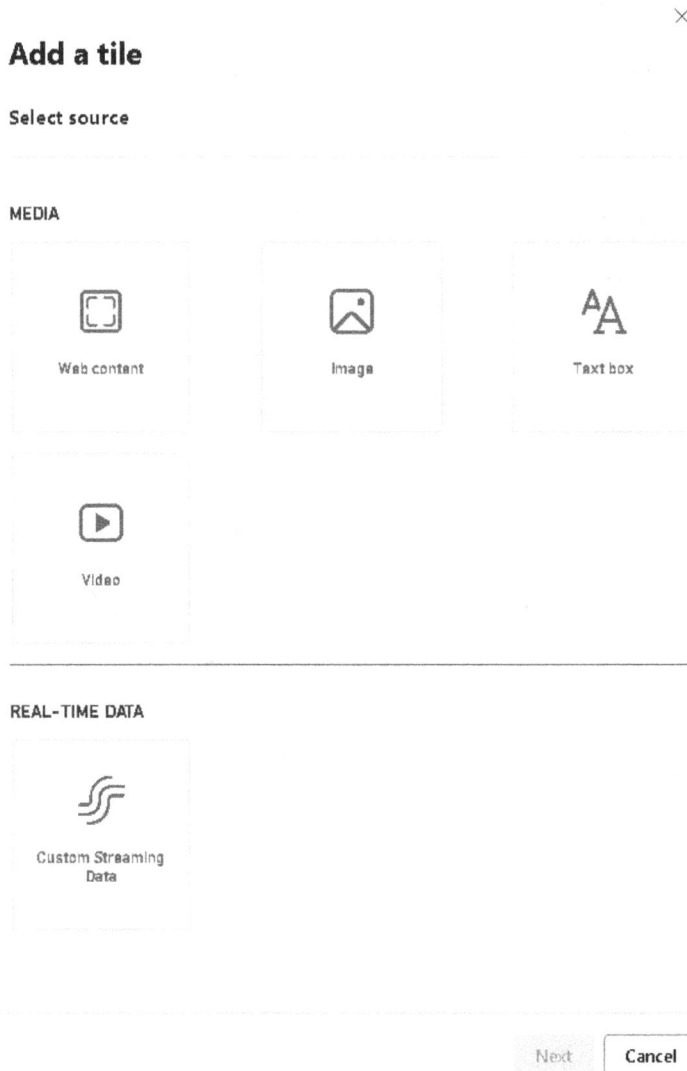

Figure 10.5 – Choosing tiles for a dashboard

5. Once you select the **Custom Streaming Data** option, you have the option of choosing an API, Azure Stream, or a PubNub streaming dataset.

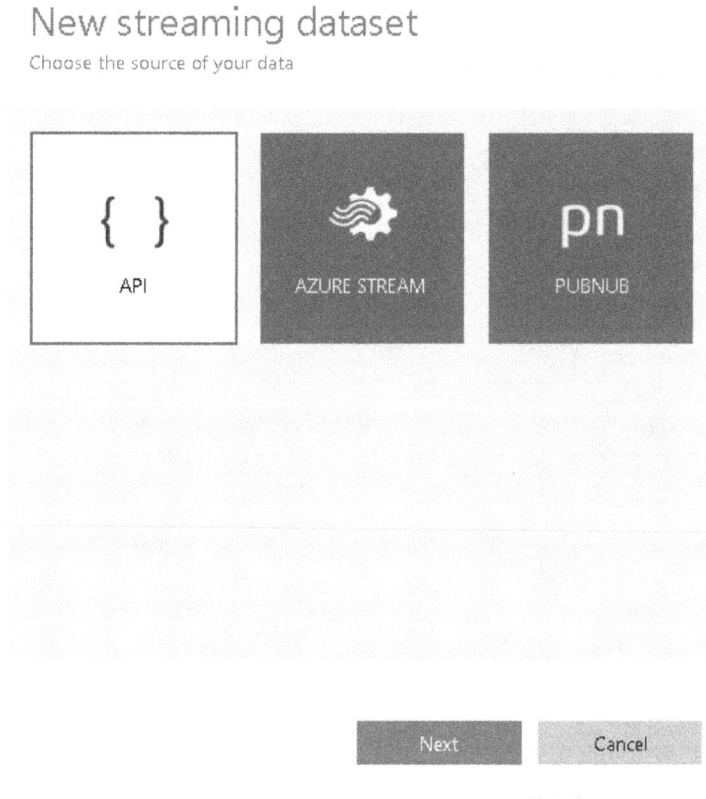

Figure 10.6 – Choosing the type of streaming data

6. After choosing the **PUBNUB** option, you need to subscribe to a PubNub stream. PubNub provides some free demo streams to use. After entering the subscription information, we can see the schema of the data.

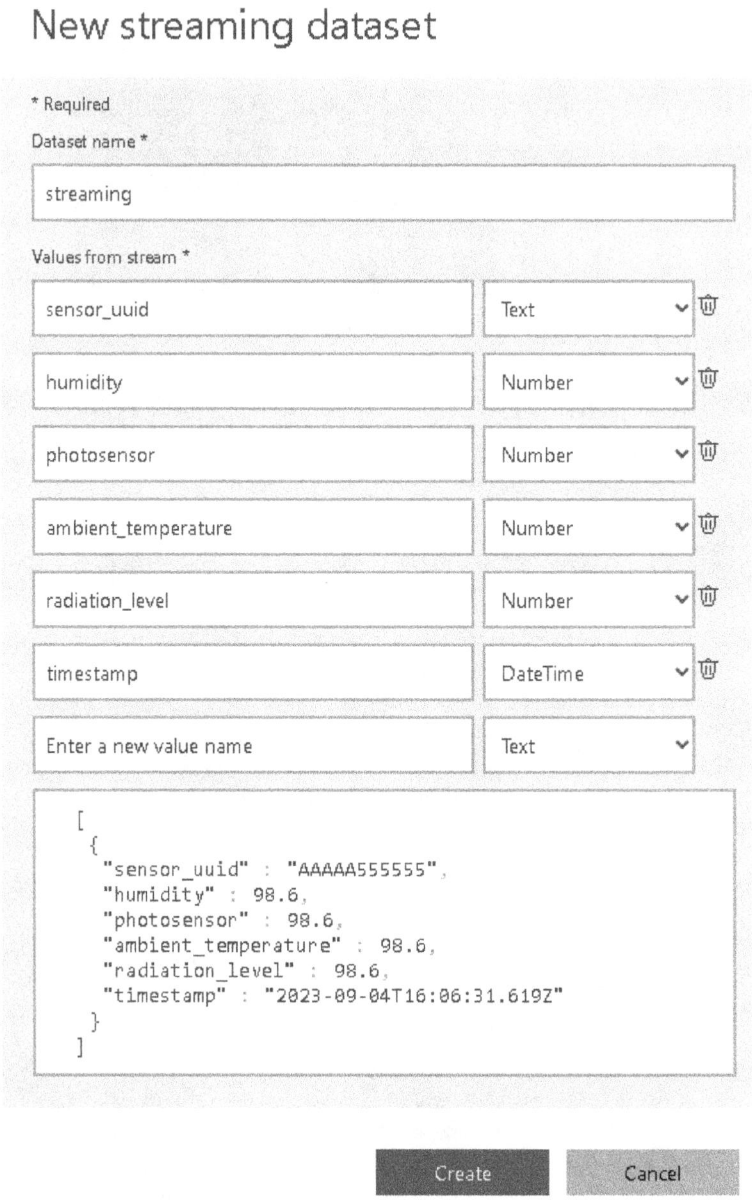

New streaming dataset

* Required

Dataset name *

| streaming |

Values from stream *

sensor_uuid		Text	
humidity		Number	
photosensor		Number	
ambient_temperature		Number	
radiation_level		Number	
timestamp		DateTime	
Enter a new value name		Text	

```
[
  {
    "sensor_uuid" : "AAAAA555555",
    "humidity" : 98.6,
    "photosensor" : 98.6,
    "ambient_temperature" : 98.6,
    "radiation_level" : 98.6,
    "timestamp" : "2023-09-04T16:06:31.619Z"
  }
]
```

Create   Cancel

Figure 10.7 – The schema of the streaming data

7. Once you create the dataset, you can choose to add a tile again and select the dataset from the list of datasets created. Now, we can choose the type of visual and the data to display.

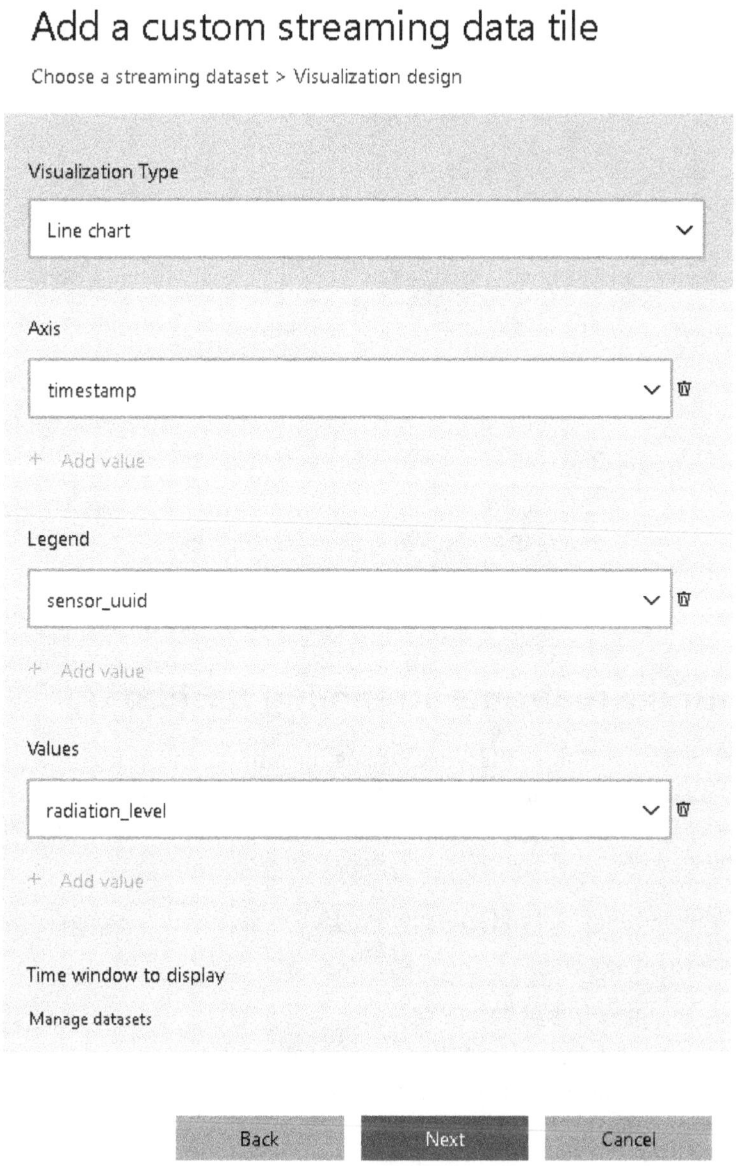

Add a custom streaming data tile

Choose a streaming dataset > Visualization design

Visualization Type

Line chart

Axis

timestamp

+ Add value

Legend

sensor_uuid

+ Add value

Values

radiation_level

+ Add value

Time window to display

Manage datasets

Back   Next   Cancel

Figure 10.8 – Setting up the visual to display

After a slight delay, you will start to see real-time data flowing across the tile.

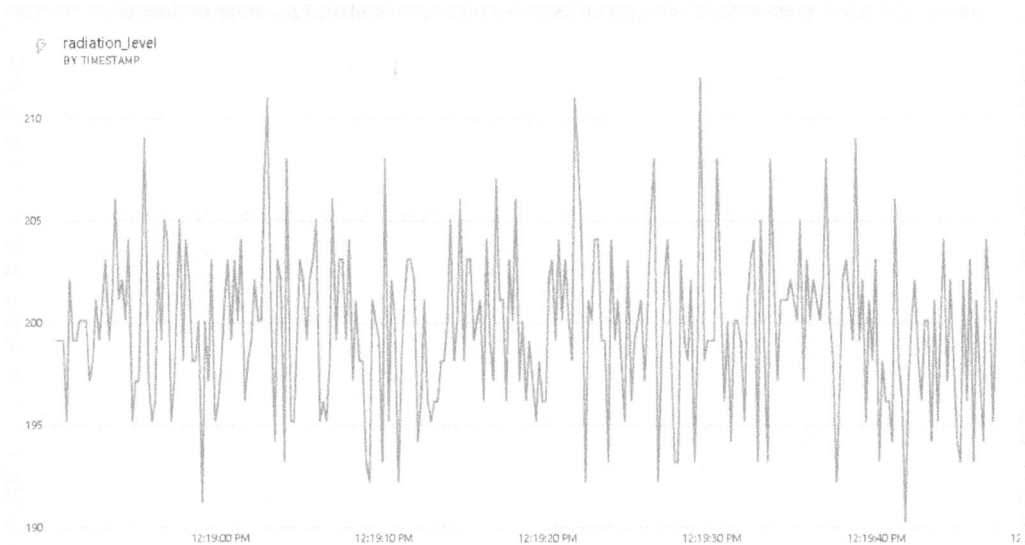

Figure 10.9 – Visualizing the streaming data

Now that you have seen how to set up real-time data visuals on a dashboard, it is time to try it yourself.

Lab – Creating a real-time streaming dashboard

In this lab, you will create a real-time data visual on a dashboard. You will be using a simulated sensor that sends data to Azure Analytics and from there to the dashboard. To complete this lab, make sure you have a Power BI Pro license (you can choose either a paid subscription or a free trial). So, let's get started with the lab:

1. Create a resource group, and add an Azure IoT Hub to the resource group.

2. Add a device to the IoT Hub.

3. Copy the connection string of the device.

4. Using the simulator at `https://azure-samples.github.io/raspberry-pi-web-simulator`, add the connection string on *line 15* and start the simulator.

5. Check your IoT hub, and make sure it is getting signals.

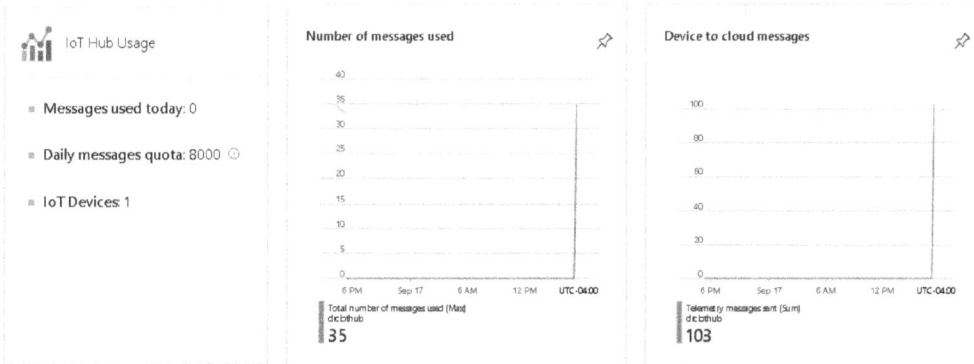

Figure 10.10 – Checking the IoT Hub for signals received

6. You can stop the simulator.

7. Log in to Power BI and create a new workspace called `streamingdemo`.

8. Select the `https://app.powerbi.com/groups/me/list?experience=kusto` workspace. This will contain the GUID of the workspace. If you are asked to log in, use the account you used to sign up for Power BI.

9. Create an Azure Stream Analytics job in the resource group.

10. For the input add the IoT Hub.

11. For the output, select **Power BI**, and enter the information needed to connect to the workspace. For the group workspace, enter the GUID that was contained in the URL you copied in *step 7*. Don't forget to select **Authorize** to log into your Power BI workspace. (A separate web page will pop up.)

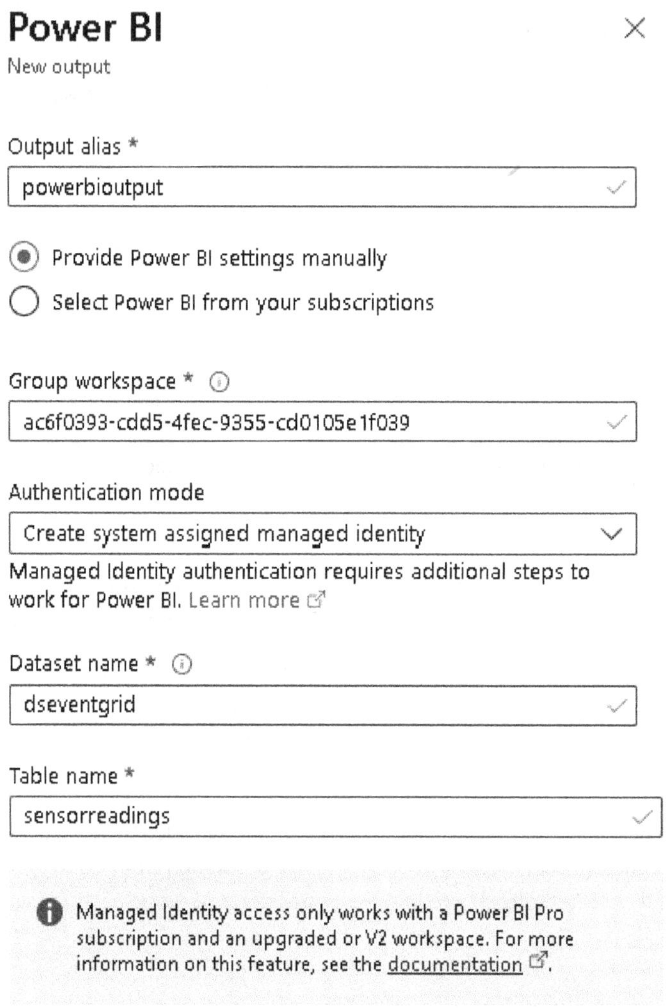

Power BI
New output   ✕

Output alias *

```
powerbioutput                                    ✓
```

◉ Provide Power BI settings manually

◯ Select Power BI from your subscriptions

Group workspace * ⓘ

```
ac6f0393-cdd5-4fec-9355-cd0105e1f039             ✓
```

Authentication mode

```
Create system assigned managed identity          ⌄
```

Managed Identity authentication requires additional steps to work for Power BI. Learn more ⌐ʾ

Dataset name * ⓘ

```
dseventgrid                                      ✓
```

Table name *

```
sensorreadings                                   ✓
```

ⓘ Managed Identity access only works with a Power BI Pro subscription and an upgraded or V2 workspace. For more information on this feature, see the <u>documentation</u> ⌐ʾ.

Figure 10.11 – Adding a Power BI output

12. For the query, enter the following, and update the input alias and output alias with your values:

```
SELECT
    *
INTO
    [YourOutputAlias]
FROM
    [YourInputAlias]
```

13. Start the simulator, and verify that it is sending signals. You will see them in the output window.

14. You can start the job by clicking the **Start** button at the top of the page.

15. In Power BI, go to the workspace, choose **New**, and then **Dashboard**.

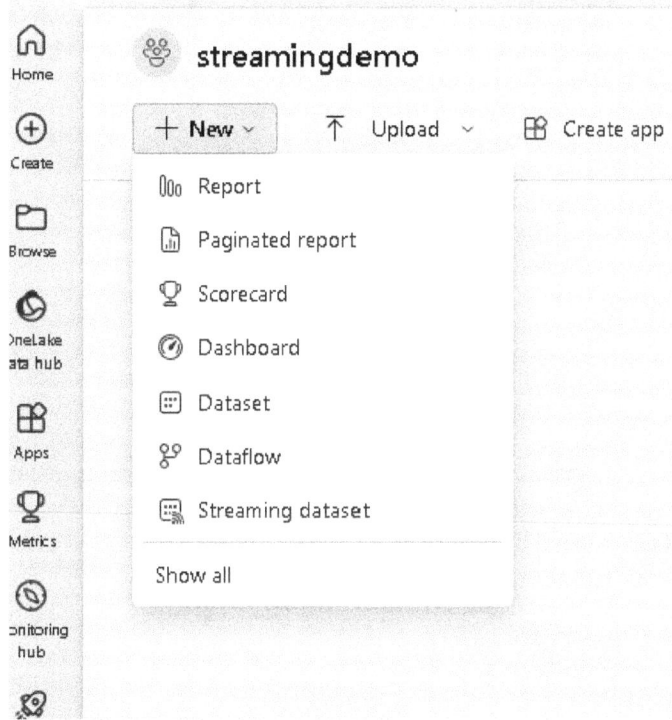

Figure 10.12 – Adding a dashboard

16. After creating the dashboard, select **Add a tile** from the **Edit** dropdown.

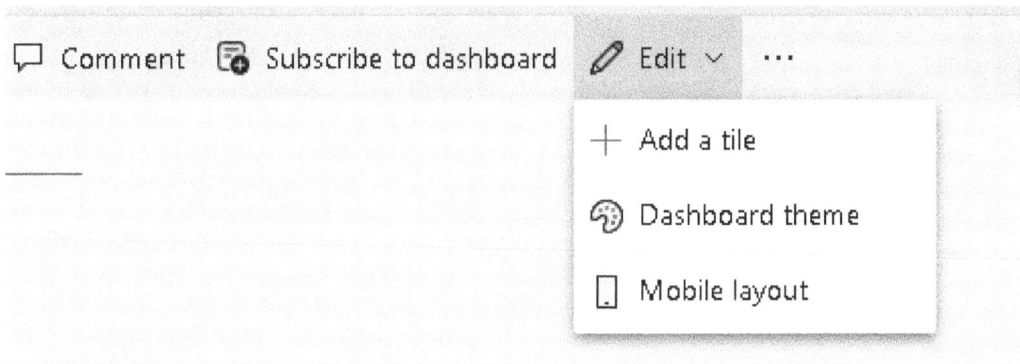

Figure 10.13 – Adding a tile

17. On the **Add a tile** form, select **Custom Streaming Data**.

18. On the next screen, you should see your dataset.

Add a custom streaming data tile

Choose a streaming dataset

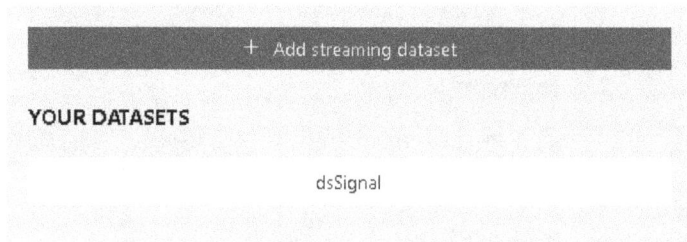

```
              +  Add streaming dataset
```

YOUR DATASETS

dsSignal

Figure 10.14 – Adding your streaming dataset

19. Next, select **Card** for **Visualization Type**, and then select the **temperature** field.

Add a custom streaming data tile

Choose a streaming dataset > Visualization design

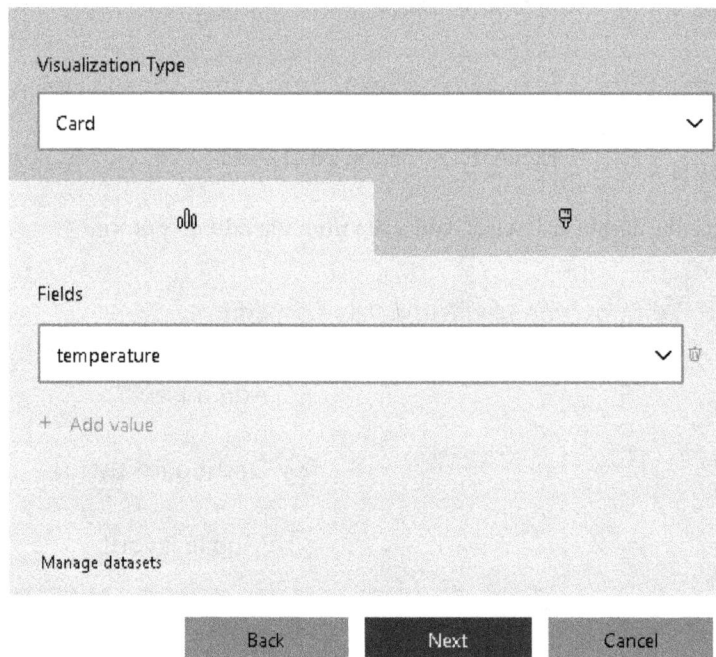

Visualization Type

```
Card                                              ∨
```

Fields

```
temperature                                    ∨  🗑
```

+ Add value

Manage datasets

```
        Back              Next              Cancel
```

Figure 10.15 – Creating a card for temperature

20. Try creating a line chart for humidity.

21. When done, clean up your resources.

Now that you have completed the lab and gained some hands-on experience displaying real-time data, let us summarize what we covered in this chapter and what is coming next.

Summary

In this chapter, we delved into the fundamentals of Power BI Dashboard creation, focusing on real-time data integration and visualization. We began by exploring the Power BI dashboard basics and understanding the core concepts. Then, we dived into the world of real-time datasets, where we discussed the various types of data sources that can be leveraged to create dynamic dashboards. Whether it's streaming data from IoT devices, databases, or external APIs, we examined the diverse options available for real-time data integration.

Moving forward, we ventured into the art of displaying real-time visuals. We covered dynamic charts and live tiles, ensuring that our dashboards provide users with up-to-the-minute information.

To solidify our understanding and put theory into practice, we embarked on a hands-on lab – *Creating a real-time streaming dashboard*. Through step-by-step guidance, we applied the knowledge acquired throughout the chapter to build a real-time streaming dashboard from scratch, utilizing Power BI's features and capabilities.

By now, you should have a firm grasp of the Power BI dashboard basics, be well-versed in the types of real-time datasets, understand the process of pushing data into datasets, and be proficient in displaying real-time visuals. The practical experience gained through the lab exercise ensures that you are equipped to create compelling, real-time dashboards that empower data-driven decision-making.

In the next chapter, we will look at machine learning. Machine learning is the arrangement of mathematical operations that process data input into actionable output. You can incorporate machine learning into your IoT solution to help analyze and gain insight into the data. This can be done in the cloud or on the edge. We will cover use cases for each, determining when to use cloud compute and when to use the compute on the edge device.

11

Integrating Machine Learning

In today's digital age, the fusion of **machine learning** (**ML**) and IoT technologies has paved the way for groundbreaking innovations across various industries. This chapter delves into the fundamental concepts that underpin this powerful combination, offering you a comprehensive understanding of ML.

We will embark on a journey through the realms of data-driven decision-making, exploring how the seamless integration of these technologies can revolutionize the way we approach real-world challenges. With hands-on experience in mind, we'll also walk through a practical laboratory exercise, where you will have the opportunity to create a predictive maintenance system—an excellent showcase of how the synergy between ML and IoT can transform businesses and drive innovation.

In this chapter, we will cover the following topics:

- Understanding ML basics
- What are Azure **artificial intelligence** (**AI**) and ML services?
- Exploring ML on the edge
- Combining IoT with ML
- Lab – creating an anomaly detection system

Understanding ML basics

Although AI and ML are often used interchangeably, there is a distinction. AI refers to the simulation of human intelligence in machines or computer systems. It is a multidisciplinary field of computer science and engineering that aims to create intelligent agents or systems capable of perceiving their environment, reasoning, learning from experience, and making decisions to achieve specific goals.

Key components and concepts of AI include:

- **ML**: ML is a subset of AI that focuses on the development of algorithms and models that allow machines to learn from data and make predictions or decisions without being explicitly programmed. It is a fundamental tool in AI research and applications.

- **Neural networks (NNs)**: NNs are computational models inspired by the human brain's structure and functioning. **Deep learning (DL)**, a subset of ML, utilizes **deep NNs (DNNs)** with multiple layers to handle complex tasks such as image and speech recognition.

- **Natural language processing (NLP)**: NLP is a subfield of AI that deals with the interaction between computers and human language. It enables machines to understand, interpret, and generate human language, making applications such as chatbots and language translation possible.

- **Computer vision (CV)**: CV is a field of AI that focuses on enabling machines to interpret and understand visual information from the world, including images and videos. This technology is used in applications such as facial recognition and object detection.

- **Robotics**: AI is used in robotics to create intelligent robots and autonomous systems that can perform tasks in the physical world. These robots can range from simple industrial robots to advanced humanoid robots.

- **Expert systems**: Expert systems are AI programs designed to mimic the decision-making abilities of a human expert in a particular domain. They use knowledge-based rules and logic to solve complex problems.

- **Reinforcement learning (RL)**: This is a type of ML where agents learn to make a sequence of decisions by interacting with an environment. RL is used in applications such as game playing and autonomous control.

- **AI ethics and bias**: As AI technologies become more prevalent, ethical considerations regarding their use, fairness, transparency, and accountability have gained significant attention. Addressing bias in AI algorithms and ensuring responsible AI development is an important aspect of AI research and deployment.

AI has a broad range of applications across various industries, including healthcare, finance, transportation, entertainment, and more. Some common examples of AI applications include virtual personal assistants (for example, Siri, Alexa), recommendation systems (for example, Netflix recommendations), autonomous vehicles, medical diagnosis and treatment planning, and fraud detection.

AI continues to advance rapidly, with ongoing research in areas such as explainable AI, AI safety, and AI ethics. Its potential to transform industries and improve the quality of life for individuals is significant, but it also raises important questions about the responsible development and deployment of AI technologies.

ML is a subfield of AI that focuses on developing algorithms and models that enable computers to learn and make predictions or decisions without being explicitly programmed to do so. It is a data-driven approach to solving complex problems and is based on the idea that computers can automatically improve their performance on a specific task through the analysis of data.

Here are some key concepts and components of ML:

- **Data**: ML relies heavily on data. Algorithms are trained on large datasets that contain examples of input-output pairs or patterns. This data is used to learn underlying relationships and patterns that can be used for prediction or classification.

- **Training**: During the training phase, an ML model is exposed to the training data, and it adjusts its internal parameters to minimize the error or difference between its predictions and the actual outcomes in the training dataset. This process involves optimization techniques to fine-tune the model.

- **Features**: Features are the input variables or attributes that the model uses to make predictions or classifications. The selection and engineering of relevant features can greatly impact the performance of an ML model.

- **Algorithms**: ML algorithms are mathematical and statistical models that define how a model learns from data and makes predictions. Common types of ML algorithms include linear regression, decision trees, **support vector machines** (**SVMs**), and NNs.

- **Supervised learning (SL) versus unsupervised learning (UL)**: In SL, the algorithm is trained on labeled data, where each example has a known output. The model learns to map inputs to outputs. In UL, the algorithm is trained on unlabeled data, and it aims to discover hidden patterns or structures in the data.

- **Evaluation and testing**: After training, the ML model is evaluated using a separate dataset, often called a test set, to assess its performance and generalization to new, unseen data. Common evaluation metrics include accuracy, precision, recall, and F1-score, among others.

- **Overfitting and underfitting**: Overfitting occurs when a model learns the training data too well and performs poorly on new data because it has essentially memorized the training examples. Underfitting, on the other hand, happens when a model is too simple to capture the underlying patterns in the data and also performs poorly.

- **DL**: DL is a subset of ML that focuses on NNs with multiple layers (DNNs). It has been particularly successful in tasks such as image and speech recognition, NLP, and game playing.

ML has a wide range of applications, including image and speech recognition, NLP, recommendation systems, autonomous vehicles, medical diagnosis, financial fraud detection, and more. These are just a few examples, and the applications of ML are continually expanding into new domains as the technology evolves. ML has the potential to optimize processes, improve decision-making, enhance user experiences, and provide valuable insights in various fields.

Now that you know why ML is becoming increasingly important, let's take a brief look at how ML algorithms are created in Azure.

Creating an ML algorithm in Azure

Although you may not create the ML algorithms you use, Microsoft has created a set of tools that make it a lot easier to construct and use ML algorithms.

Creating an ML algorithm in Azure involves using **Azure Machine Learning** (**Azure ML**), a cloud-based platform that provides tools, services, and resources for building, training, deploying, and managing ML models.

Here's a general outline of the steps to create an ML algorithm in Azure:

1. **Create an Azure ML workspace**: The **workspace** is the top-level resource for your ML activities, providing a centralized place to view and manage the artifacts you create when you use Azure ML:

    I.   Sign in to Azure ML Studio: `https://ml.azure.com/home?tid=93ae660e-4895-4413-a75e-c5a4de429c0e`.

    II.   Select **Create new workspace.**

    III.   Configure the workspace settings, such as **Subscription**, **Resource group**, and **Region**, as shown in *Figure 11.1*:

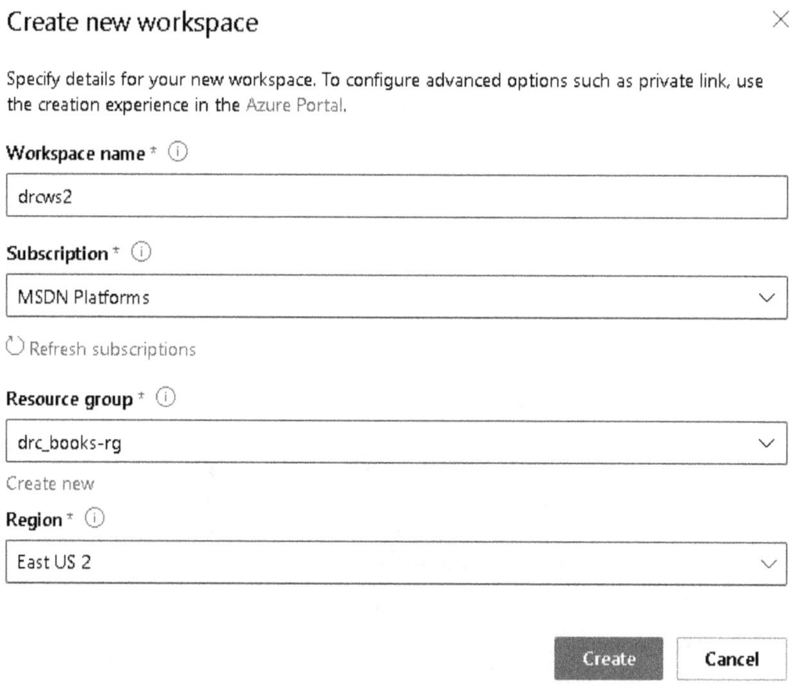

Create new workspace   ✕

Specify details for your new workspace. To configure advanced options such as private link, use the creation experience in the Azure Portal.

Workspace name * ⓘ

    drcws2

Subscription * ⓘ

    MSDN Platforms   ⌄

↻ Refresh subscriptions

Resource group * ⓘ

    drc_books-rg   ⌄

Create new

Region * ⓘ

    East US 2   ⌄

                                    [Create]   [Cancel]

Figure 11.1 – Creating an ML workspace

2. **Data preparation**:

   I.   Upload your dataset to Azure, using either Azure Blob Storage, Azure SQL Database, or other data storage options.

   II.   Explore and preprocess your data using tools such as Jupyter Notebook or Azure Data Studio.

The following screenshot shows a section of the Azure ML designer. The first box is for retrieving the data. The second box is for selecting the fields. Notice the selection criteria have excluded a column that has many missing values. The final box is for removing rows with many missing values:

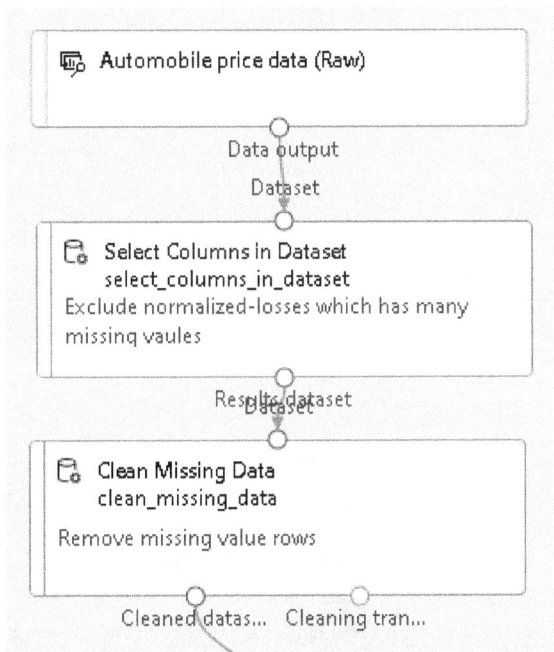

Figure 11.2 – Cleaning and prepping the data

3. **Create an ML experiment**:

   I.   Use Azure ML Studio to create a new experiment.

   II.   Define the experiment's objectives, and select the algorithm or model you want to use.

4. **Data splitting**: Split your dataset into training, validation, and test sets. This is essential for evaluating the model's performance accurately.

5. **Feature engineering**: Perform feature selection and engineering to prepare the data for training. Azure provides data preprocessing capabilities to help with this.

6. **Model training**:

    I.   Select the ML algorithm or model you want to use (for example, scikit-learn, TensorFlow, or PyTorch).

    II. Configure the training parameters and train the model on your training data using Azure ML's training cluster.

The following screenshot shows another section of the Azure ML designer.

This box is running an ML algorithm using a Python script that predicts customer churn:

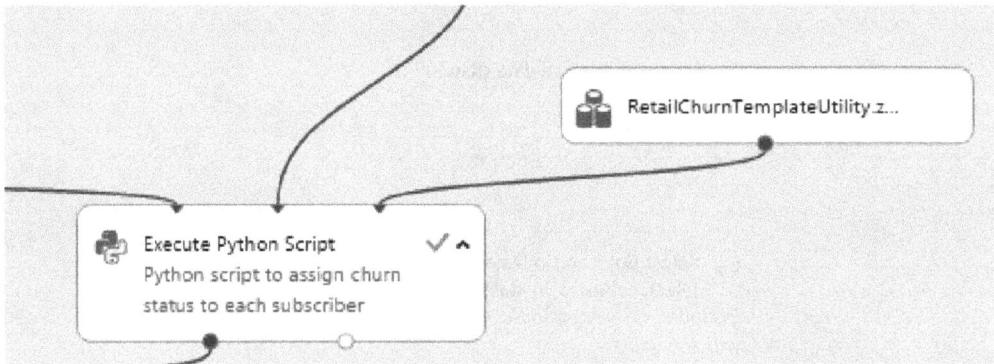

Figure 11.3 – Training the model

7. **Hyperparameter tuning** (*optional*): Hyperparameter tuning, also known as hyperparameter optimization or hyperparameter search, is a crucial step in the ML model development process. Hyperparameters are settings or configurations for ML algorithms that are not learned from the data but must be set prior to training the model. These settings significantly impact the model's performance and generalization abilities and depend on the model you are using. Not all models need hyperparameters. Common examples of hyperparameters include learning rates, the number of hidden layers and units in an NN, the depth of decision trees, regularization strength, batch size, and more.

8. **Model evaluation**: Evaluate your trained model's performance using the validation dataset. Azure provides tools for generating evaluation metrics such as accuracy, precision, recall, and more.

9. **Model deployment**: Once you are satisfied with your model's performance, deploy it as a web service or container. Azure ML supports various deployment options, including **Azure Kubernetes Service (AKS)** and **Azure Container Instances (ACI)**.

10. **Monitoring and management**: Monitor your deployed model's performance and usage through Azure ML's monitoring capabilities. You can also manage and update your model as needed.

11. **Scalability and optimization**: Depending on your requirements, you can scale your model's deployment to handle increased workloads efficiently.

12. **Testing and Continuous Integration/Continuous Deployment (CI/CD)**: Implement CI/CD pipelines to automate the testing and deployment of new model versions.

13. **Security and compliance**: Ensure that your ML solution complies with security and privacy regulations. Azure provides tools and features to help with compliance.

14. **Logging and error handling**: Implement logging and error-handling mechanisms to track and diagnose issues in your ML system.

15. **Maintenance and updates**: Regularly monitor and maintain your deployed models to ensure they remain accurate and effective over time.

Azure provides a comprehensive set of tools and services to streamline the end-to-end ML workflow, from data preparation to model deployment and management. Azure ML Studio, Azure Databricks, and Azure Notebooks are some of the services and tools commonly used in Azure's ML ecosystem. Additionally, Azure DevOps can be used for automating CI/CD pipelines for ML projects.

Although you just got a taste of what creating an ML algorithm is like, using the Azure toolset, you probably will not be creating them but rather, will be using one created for you by a data scientist. If you want to learn more about the ML process, a great place to start is this web page: `https://azure.microsoft.com/en-us/products/machine-learning/`.

Now that we have a basic understanding of ML algorithms in Azure, let us learn about the various AI and ML services that one can use.

What are Azure AI and ML services?

Azure AI (Cognitive) Services is a suite of cloud-based AI and ML services provided by Microsoft Azure. These services enable developers to integrate AI capabilities into their applications, making them more intelligent and capable of understanding, processing, and responding to human language, images, and other forms of data. Azure Cognitive Services simplifies the integration of AI technologies, allowing developers to focus on building innovative solutions without the need for extensive AI expertise.

Here are some key features and capabilities of Azure AI (Cognitive) Services:

- **Speech services**: Azure offers a range of speech-related services, including the following:

  - **Speech-to-Text**: Convert spoken language into written text for transcription, voice assistants, and more

  - **Text-to-Speech**: Generate natural-sounding speech from text for applications such as chatbots and voice interfaces

  - **Speaker Recognition**: Identify and verify individuals based on their voice characteristics

- **Language services**: These services are designed to understand and process human language:

  - **Text Analytics**: Analyze text data for **sentiment analysis (SA)**, keyphrase extraction, and language detection

  - **Language Understanding Intelligent Service (LUIS)**: Build **natural language understanding (NLU)** models to interpret user intents from text

  - **Translator Text**: Translate text between languages in real time

- **Vision services**: Azure Cognitive Services includes CV capabilities:

  - **Computer Vision**: Analyze and interpret images and videos to extract information, detect objects, and recognize text

  - **Face API**: Detect, identify, and analyze faces in images and videos

  - **Custom Vision**: Train custom image classification and object detection models

- **Knowledge mining**: Azure Cognitive Search enables you to explore and extract insights from structured and unstructured data, making it easier to search and discover information within your documents and databases

- **Decision services**: These services help in making data-driven decisions:

  - **Anomaly Detector**: Detect anomalies and irregular patterns in time-series data

  - **Personalizer**: Build RL models to optimize content recommendations and user experiences

- **Content Moderator**: A service that helps identify and moderate potentially inappropriate or offensive content in text, images, and videos

- **Form Recognizer**: Extract structured data from forms, invoices, and receipts using **optical character recognition (OCR)**

- **Ink Recognizer**: Recognize and convert handwritten ink strokes into digital text and shapes

- **QnA Maker**: Create and manage a knowledge base for frequently asked questions and integrate it into chatbots and applications

As you can see, there are many AI services, and more are being developed. Gone are the days of needing a full-time department of data scientists unless you need to develop proprietary, custom algorithms.

Azure AI (Cognitive) Services is designed to be accessible through REST APIs, **software development kits (SDKs)**, and other developer-friendly interfaces, making it straightforward to integrate AI capabilities into various applications, including web and mobile apps, chatbots, and IoT devices. These services are backed by Microsoft's extensive research and development in AI, ensuring high-quality and reliable AI-powered solutions. Developers can leverage these services to enhance user experiences, improve decision-making, and automate various tasks using AI-driven insights and intelligence.

One example of a useful AI service is anomaly detection, which is a common use case for a lot of enterprises. **Anomaly detection**, an AI methodology, is employed to ascertain whether data points within a sequence conform to anticipated criteria.

The utility of anomaly detection spans various contexts. To illustrate, a sophisticated HVAC system might employ anomaly detection to oversee building temperatures and trigger an alarm if they deviate from the expected range for a specified duration. Since this is a common use case, there is a set of common algorithms used for different types of anomalies. Microsoft can supply anomaly detection, so you do not have to roll your own.

Now that you know about the vast options of out-of-the-box AI services provided by Microsoft, let us look at why we would want to move them to an edge device.

Exploring ML on the edge

Moving ML to an edge device has several advantages and use cases, depending on the specific requirements of a project or application. Here are some reasons why you might consider deploying ML to edge devices:

- **Low latency and real-time processing**: Edge devices are located close to the data source or the point of action, which reduces the time it takes for data to travel to a centralized server or cloud. This proximity allows for real-time processing, making it suitable for applications where low latency is critical, such as autonomous vehicles, industrial automation, and robotics.

- **Privacy and data security**: Edge computing allows sensitive data to be processed locally on the device, rather than sending it to a remote server or cloud. This can enhance privacy and data security by reducing the risk of data breaches during transit and storage. It's especially important in applications such as healthcare, finance, and surveillance.

- **Bandwidth efficiency**: Edge devices can preprocess and filter data before transmitting it to the cloud, reducing the volume of data that needs to be sent over the network. This is particularly valuable in situations with limited bandwidth or high data transfer costs.

- **Offline and disconnected operation**: Edge devices can continue to function and make decisions even when they are not connected to the internet or a central server. This capability is beneficial in remote locations, on mobile devices, or in scenarios where network connectivity is unreliable.

- **Improved reliability**: Decentralized processing at the edge can enhance system reliability because it reduces dependency on a **single point of failure** (**SPOF**), such as a central server or cloud service. Failures in one edge device typically don't affect the operation of others.

- **Compliance and regulation**: Some industries and regions have strict regulations governing data storage and processing. Edge computing allows organizations to comply with these regulations by keeping data within specific geographic boundaries or on premises.

- **Cost efficiency**: Edge devices can help optimize cloud usage and reduce costs. Instead of sending all data to the cloud for processing, only relevant or anomalous data may be sent, resulting in potential cost savings for data transfer and cloud processing.

- **Scalability and distributed processing**: Edge computing enables distributed processing across multiple edge devices, making it easier to scale horizontally. You can add more edge devices as needed to handle increased workloads.

- **Improved user experience**: Applications that require immediate responses, such as **augmented reality** (**AR**) and **virtual reality** (**VR**) applications, benefit from edge-based ML models that can provide real-time feedback without relying on distant servers.

- **Offline ML**: Edge devices can host pre-trained ML models, allowing them to make predictions and decisions without requiring constant internet connectivity. This is useful in applications such as recommendation systems on smartphones or IoT devices.

While there are numerous advantages to moving ML to edge devices, it's essential to consider the trade-offs, including limited computational resources and potential challenges in managing and updating models across a distributed edge network. The choice to use edge computing should align with the specific requirements and constraints of your project or application.

If you do decide to run AI on the edge, you have to add AI modules on the edge device. To do this, you first create an edge device in the IoT hub. You can then add modules to the device by using the **Set modules on device** page, as shown in the following screenshot:

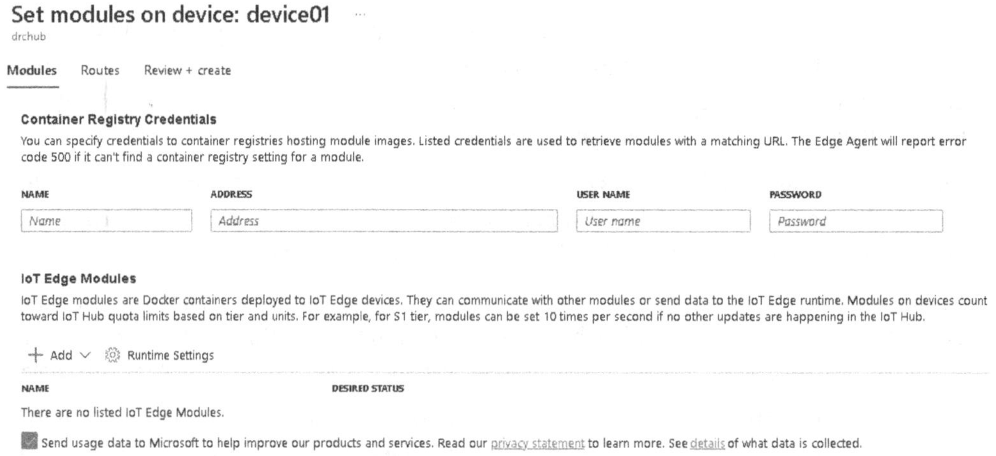

Set modules on device: device01
drchub

Modules   Routes   Review + create

Container Registry Credentials

You can specify credentials to container registries hosting module images. Listed credentials are used to retrieve modules with a matching URL. The Edge Agent will report error code 500 if it can't find a container registry setting for a module.

NAME	ADDRESS	USER NAME	PASSWORD
Name	Address	User name	Password

IoT Edge Modules

IoT Edge modules are Docker containers deployed to IoT Edge devices. They can communicate with other modules or send data to the IoT Edge runtime. Modules on devices count toward IoT Hub quota limits based on tier and units. For example, for S1 tier, modules can be set 10 times per second if no other updates are happening in the IoT Hub.

+ Add ⌄   ⚙ Runtime Settings

NAME	DESIRED STATUS

There are no listed IoT Edge Modules.

☑ Send usage data to Microsoft to help improve our products and services. Read our privacy statement to learn more. See details of what data is collected.

Figure 11.4 – Adding an edge module

The modules are Docker containers hosted on a container registry. You can create your own but are more likely to use one that is available to you on the marketplace. There are many modules to choose from, including ones created by Microsoft and third parties, as seen in *Figure 11.5*:

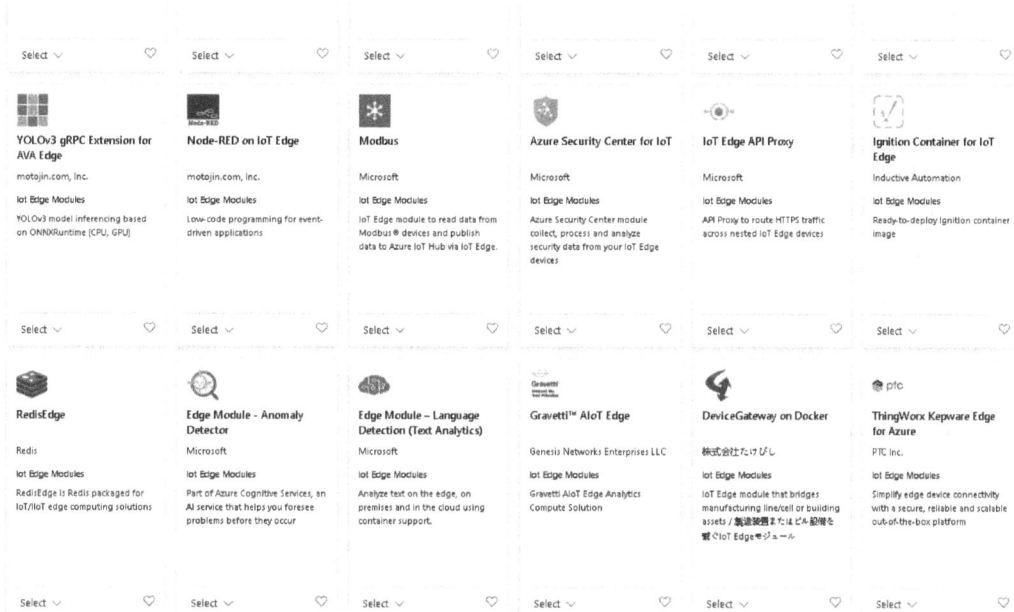

Figure 11.5 – Adding a module from the marketplace

Next, let's take a look at some of the use cases we can solve by combining IoT with ML.

Combining IoT with ML

Combining IoT with ML opens a wide range of innovative use cases across various industries. Here are some common and impactful use cases for IoT and ML integration:

- **Predictive maintenance**: Sensors on machinery and equipment collect data on factors such as temperature, vibration, and wear. ML models analyze this data to predict when maintenance is needed, reducing downtime and preventing costly breakdowns.

  Industry: Manufacturing, energy, and transportation

- **Anomaly detection**: IoT devices monitor data streams, such as patient vitals, financial transactions, or network traffic. ML algorithms detect anomalies and raise alerts for potential issues or security breaches.

  Industry: Healthcare, finance, and security

- **Smart agriculture**: In the domain of agriculture, IoT sensors and cameras in fields gather data on soil conditions, weather, crop health, and animal behavior. ML models provide insights into optimal planting times, irrigation, and pest control.

- **Energy management**: Smart meters and sensors measure energy consumption in real time. ML algorithms analyze this data to optimize energy usage, predict demand, and reduce costs.

  Industry: Commercial and residential

- **Environmental monitoring**: IoT devices collect data on air quality, water quality, temperature, and wildlife movement. ML models help in analyzing and predicting environmental changes and their impact.

  Industry: Environmental science and conservation

- **Supply chain optimization**: IoT sensors track the location, condition, and status of goods in transit. ML algorithms optimize routing, predict delivery times, and manage inventory more efficiently.

  Industry: Logistics and retail

- **Healthcare wearables**: Within the realm of healthcare, wearable IoT devices, such as fitness trackers and medical monitors, collect data on user health. ML models analyze this data to provide personalized health recommendations and detect health issues.

- **Smart cities**: In urban planning, IoT sensors and cameras monitor traffic, pollution levels, waste management, and public safety. ML helps in traffic optimization, crime prediction, and resource allocation.

- **Retail customer insights**: In retail, IoT beacons and cameras in stores collect data on customer behavior. ML algorithms analyze this data to provide personalized product recommendations and optimize store layouts.

- **Fleet management**: IoT devices in vehicles monitor driver behavior, fuel consumption, and vehicle health. ML models optimize routes, improve driver safety, and reduce fuel costs.

  Industry: Transportation and logistics

- **Building automation**: In commercial real estate, IoT sensors control lighting, HVAC systems, and security in buildings. ML models optimize energy usage, enhance security, and improve comfort for occupants.

- **Oil and gas industry**: In the oil and gas sector, IoT sensors on drilling rigs and pipelines monitor equipment status and environmental conditions. ML models predict equipment failures and optimize operations.

- **Quality control in manufacturing**: IoT sensors on production lines collect data on product quality. ML models identify defects and deviations, ensuring consistent quality.

- **Water management**: IoT devices monitor water levels, quality, and distribution. ML models optimize water resource allocation and predict water shortages.

  Industry: Utilities and agriculture

- **Home automation**: In the residential industry, IoT devices control lighting, thermostats, and appliances in smart homes. ML learns user preferences and adjusts settings for energy efficiency and comfort.

These are just a few examples of how combining IoT and ML can create innovative solutions across diverse domains. The ability to collect and analyze data from IoT devices in real time or near real time enables organizations to make data-driven decisions, automate processes, and improve efficiency and customer experiences.

Now that we have discussed the basics of ML and AI, it is time to get some hands-on experience incorporated into an IoT system in the following lab.

Lab – creating an anomaly detection system

Azure Stream Analytics (**ASA**) simplifies the process of creating and training custom ML models by incorporating built-in anomaly detection powered by ML. It offers the convenience of performing anomaly detection through straightforward function calls. Two novel unsupervised ML functions have been introduced by Microsoft to identify two prevalent types of anomalies: transient and enduring. These are common anomalies, so you don't have to create your own detection algorithm but can use the ones provided by Microsoft.

The `AnomalyDetection_SpikeAndDip` function is designed to pinpoint transient or short-lived anomalies, such as spikes or dips, leveraging the widely recognized **kernel density estimation** (**KDE**) algorithm.

On the other hand, the `AnomalyDetection_ChangePoint` function is employed to identify persistent or long-lasting anomalies, such as bi-level shifts, gradual increases, and gradual decreases. It relies on the established algorithm known as exchangeability martingales. To learn more about these functions, see the following website: `https://learn.microsoft.com/en-us/stream-analytics-query/analytic-functions-azure-stream-analytics`.

In this lab, you will:

- Set up anomaly detection using ASA
- Test the anomaly detection with a simulated device

First, let's set up a simulated device sending signals to an IoT hub:

1. Log in to your Azure subscription and create a resource group and an IoT hub in a location near you.

2. Create a device and copy the connection string.

3. Go to the device simulator at `https://azure-samples.github.io/raspberry-pi-web-simulator/`.

4. Paste the connection string on *line 15* of the code in the simulated device to connect to the IoT hub.

5. Run the device for a few minutes and verify that you are getting signals passed to the IoT hub, as shown in *Figure 11.6*:

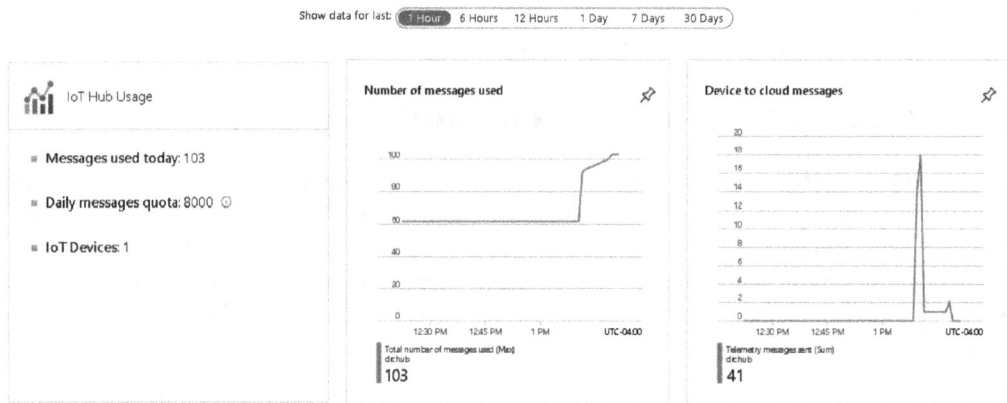

Figure 11.6 – Verifying signals

6. In the same resource group as your IoT hub, create a Stream Analytics job. Notice there is one to run on the edge and one for the cloud. Pick the one for the cloud.

7. Create a storage account in the same resource group to hold the output.

8. Once created, set your IoT hub as the input for the job.

9. If you select the **Add function** dropdown, you can select an ML service that is a published routine you created with ML Studio or select one you are still developing in the studio. You can also create Javascript **user-defined aggregates** (UDAs) or **user-defined functions** (UDFs), as depicted in *Figure 11.7*.

In this lab, we are going to use the built-in functions that are part of the ASA query language:

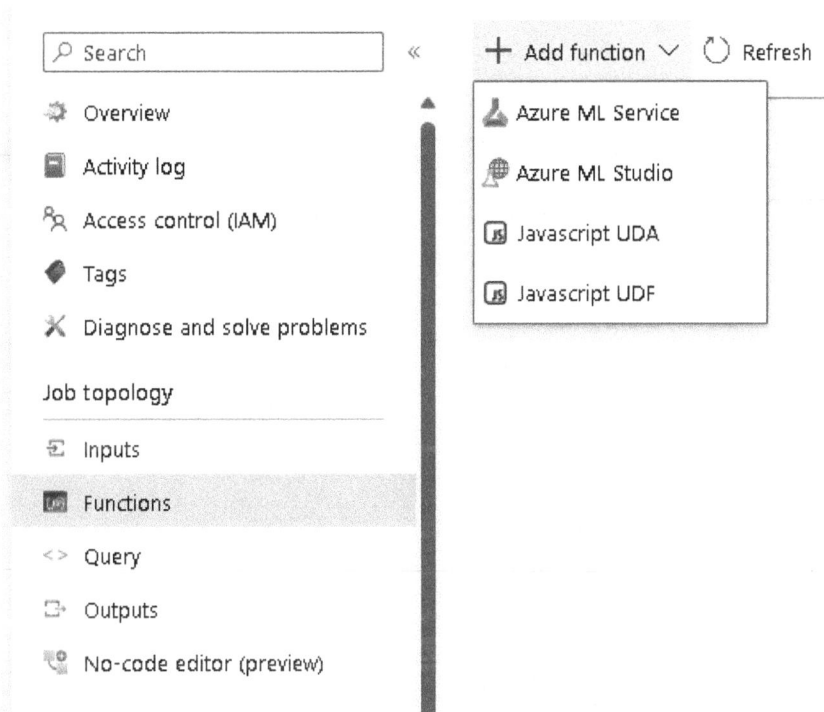

Figure 11.7 – Using the ML service

10. For the output, use the Azure storage and create a table to store the data with a **Partition key** value of deviceID and a **Row key** value of messageId. *Figure 11.8* shows this step:

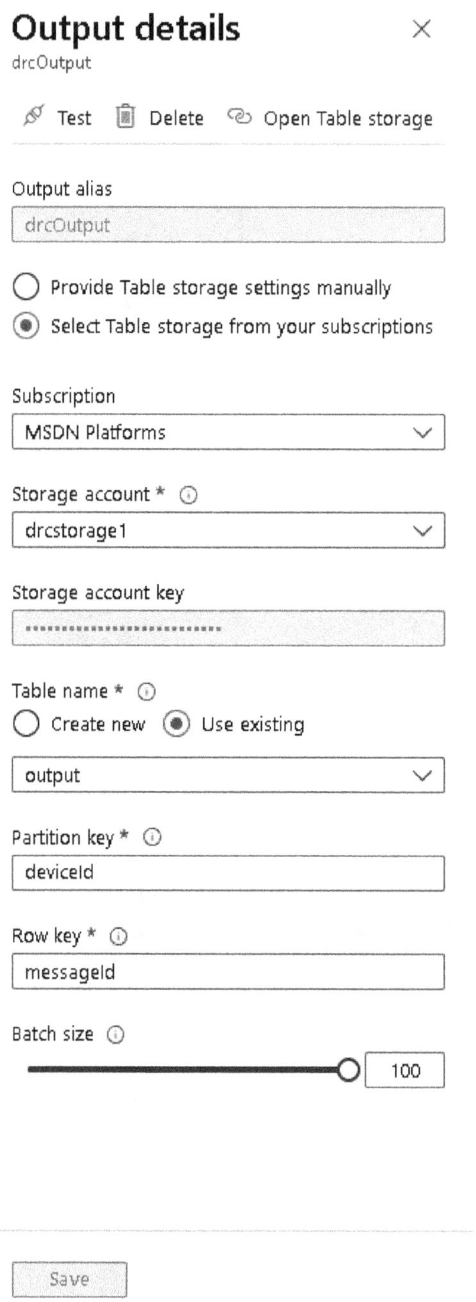

Output details

drcOutput

✐ Test   🗑 Delete   ↩ Open Table storage

Output alias

drcOutput

○ Provide Table storage settings manually
◉ Select Table storage from your subscriptions

Subscription

| MSDN Platforms | ∨ |

Storage account * ⓘ

| drcstorage1 | ∨ |

Storage account key

••••••••••••••••••••••••••••

Table name * ⓘ
○ Create new   ◉ Use existing

| output | ∨ |

Partition key * ⓘ

deviceId

Row key * ⓘ

messageId

Batch size ⓘ

─────────────────O   100

Save

Figure 11.8 – Setting up the output

11. For a query, use the following code, and replace the input and output with your values:

```
WITH AnomalyDetectionStep AS
(
    SELECT
        messageId as messageId,
        deviceId as deviceId,
        EventProcessedUtcTime AS time,
        CAST(temperature AS float) AS temp,
        AnomalyDetection_SpikeAndDip(CAST(temperature AS float),
95, 120, 'spikesanddips')
        OVER(LIMIT DURATION(second, 120)) AS SpikeAndDipScores
    FROM input
)
SELECT
    deviceId,
    messageId,
    time,
    temp,
    CAST(GetRecordPropertyValue(SpikeAndDipScores, 'Score') AS
float) AS
    SpikeAndDipScore,
    CAST(GetRecordPropertyValue(SpikeAndDipScores, 'IsAnomaly')
AS bigint) AS
    IsSpikeAndDipAnomaly
INTO output
FROM AnomalyDetectionStep
```

12. Start the simulated device and your job. Let it run for a few minutes, then stop the job and the simulator. Observe the results in the table shown in *Figure 11.9* using the storage browser in the storage account:

	PartitionKey	RowKey	Timestamp	IsSpikeAndDipAnomaly
☐	Raspberry Pi Web Client	10	2023-09-17T17:09:07.59...	0
☐	Raspberry Pi Web Client	11	2023-09-17T17:09:07.59...	0
☐	Raspberry Pi Web Client	12	2023-09-17T17:09:07.59...	0
☐	Raspberry Pi Web Client	13	2023-09-17T17:09:07.59...	1
☐	Raspberry Pi Web Client	14	2023-09-17T17:09:07.59...	1
☐	Raspberry Pi Web Client	15	2023-09-17T17:09:07.59...	0
☐	Raspberry Pi Web Client	16	2023-09-17T17:09:07.59...	0
☐	Raspberry Pi Web Client	17	2023-09-17T17:09:07.59...	1
☐	Raspberry Pi Web Client	18	2023-09-17T17:09:07.59...	0
☐	Raspberry Pi Web Client	19	2023-09-17T17:09:07.59...	0
☐	Raspberry Pi Web Client	20	2023-09-17T17:09:07.59...	0
☐	Raspberry Pi Web Client	21	2023-09-17T17:09:07.59...	0
☐	Raspberry Pi Web Client	22	2023-09-17T17:09:07.59...	0
☐	Raspberry Pi Web Client	23	2023-09-17T17:09:07.59...	0
☐	Raspberry Pi Web Client	24	2023-09-17T17:09:07.59...	0
☐	Raspberry Pi Web Client	25	2023-09-17T17:09:07.59...	0

Figure 11.9 – Observing the results of the job

13. When done, delete the resource group.

Now that you have seen how to implement anomaly detection using ASA, you may want to investigate it further. A great place to start is this web page: `https://azure.microsoft.com/en-us/blog/azure-data-explorer-and-stream-analytics-for-anomaly-detection/`. Let's sum up what we have learned and see where we are going next.

Summary

In this chapter, we embarked on a journey into the realm of data-driven decision-making, showcasing the seamless integration of technologies that can transform real-world challenges. Through hands-on experience, you created an anomaly detection system, highlighting the synergy between ML and IoT for business innovation.

Key topics covered include ML fundamentals, an introduction to Azure Cognitive Services, ML on the edge, common IoT and ML use cases, and a practical lab on building an anomaly detection system. This is one of those chapters that was an introduction to IoT and ML. If this is an area you are interested in, there is a lot more to learn. But hopefully, I have provided you with a good starting point. I do think you will most likely run into use cases where there will be an ML/AI component in an IoT system you develop during your career, and it is good knowledge to have on how to fit them together.

In the next chapter, you will learn how to react to events occurring with your IoT service and devices. To do this, you need to incorporate an Azure event grid into your IoT solution. Unlike streaming data, events are discrete and can occur randomly. These include things such as devices disconnecting or connecting, failed logins, and messages sent to an IoT hub. The next chapter will show you how to incorporate event processing into your IoT system.

12

Responding to Device Events

In the ever-expanding world of IoT, where devices are becoming increasingly interconnected and intelligent, efficient communication between these devices and systems is paramount. This chapter delves deep into the fundamental concepts of Event Grid, a powerful service that plays a pivotal role in orchestrating the flow of information within IoT ecosystems. From understanding Event Grid's core principles to mastering the art of subscribing to and responding to events, we embark on a journey that equips you with the knowledge and skills needed to manage and monitor IoT events effectively.

Our exploration begins with understanding Event Grid fundamentals, where we lay the groundwork for comprehending this essential service's architecture and components. We delve into the inner workings of Event Grid, discussing its role in facilitating seamless communication between IoT devices and the cloud.

Next, we transition to exploring common IoT events in a section dedicated to identifying and dissecting the diverse types of events that routinely occur in IoT environments. Here, we explore events such as sensor data updates, device status changes, and much more, enabling you to gain a comprehensive understanding of the event landscape you'll be navigating.

Subscribing to events is our next stop, where we dive into the mechanics of creating event subscriptions. This section guides you through the process of setting up event handlers and configuring the criteria for event delivery, ensuring that you receive only the data that matters most to your IoT solution.

The chapter then moves on to responding to events, a critical aspect of harnessing the power of Event Grid. Here, you'll learn how to design event-driven workflows, trigger actions in response to specific events, and automate processes that streamline your IoT operations.

Finally, we will wrap up with a hands-on lab on monitoring device connection and disconnection with Event Grid. In this practical exercise, you'll apply the knowledge gained throughout the chapter to create a real-world solution for monitoring the connection and disconnection of IoT devices using Event Grid. By the end of this chapter, you'll possess a solid grasp of Event Grid's core concepts, and in particular, the following main topics:

- Event Grid versus Event Hubs

- Understanding Event Grid fundamentals

- Exploring common IoT events

- Sending IoT events to Event Grid

- Responding to events

- Subscribing and responding to IoT Hub events

- Lab – conitoring device connection and disconnection with Event Grid

Event Grid versus Event Hubs

The difference between an event grid and an event hub can be confusing. Event Grid and Event Hubs are both event-driven services in Azure, but they serve different purposes and have distinct characteristics. Here are the key differences between Event Grid and Event Hubs:

- **Purpose and use cases**:

  - **Event Grid**: Event Grid is a fully managed event routing service that simplifies event-driven programming and enables you to build applications that respond to events from various Azure services and custom sources. It is designed for distributing events to multiple subscribers in near real time. Event Grid is well suited for scenarios where you want to react to events and trigger specific actions, such as sending notifications, updating data, or triggering workflows based on events.

  - **Event Hubs**: Event Hubs is a highly scalable data streaming platform and event ingestion service that is primarily used for ingesting, collecting, and processing large volumes of data in real time. It's designed for scenarios such as log and telemetry data collection, analytics, and streaming data to big data or data warehousing solutions. Event Hubs focuses on handling and managing high-throughput data streams efficiently.

- **Message types**:

  - **Event Grid**: Event Grid deals with events, which are typically lightweight notifications that something has happened. These events can include information about changes in Azure resources, custom events, or system-level events such as HTTP requests.

- **Event Hubs**: Event Hubs handles streaming data, which often consists of larger messages, such as logs, telemetry data, or other types of data that need to be processed at scale.

- **Event routing**:

  - **Event Grid**: Event Grid provides a simple and flexible event routing system. It supports filtering and routing events based on event type and custom filters, allowing you to route events to different subscribers (for example, Azure Functions, Logic Apps, or Webhooks) based on specific criteria.

  - **Event Hubs**: Event Hubs is primarily focused on ingesting and storing data streams, and it does not have built-in event routing capabilities like Event Grid. You can build custom processing pipelines to analyze and route data from Event Hubs, but it doesn't offer the same level of event-based routing out of the box.

- **Scaling**:

  - **Event Grid**: Event Grid is designed for low-latency event delivery and is suitable for scenarios where you need to react quickly to events. It automatically scales based on demand.

  - **Event Hubs**: Event Hubs is designed for handling high-throughput data streams and is optimized for scenarios where you need to process large volumes of data at scale. It supports partitioning and consumer groups for efficient data processing.

Event Grid and Event Hubs are both valuable Azure services for different use cases. Event Grid is focused on event routing and triggering actions in response to events, while Event Hubs is designed for high-throughput data streaming and analytics. Your choice between the two will depend on your specific application requirements and whether you need event-driven processing or high-throughput data streaming.

Now that you know the difference between an event grid and an event hub, let's look deeper into the Event Grid service.

Understanding Event Grid fundamentals

In the realm of IoT and **event-driven architectures** (**EDAs**), understanding the core principles of **Event Grid** is essential. Event Grid serves as the backbone for managing and orchestrating events within an ecosystem of interconnected devices and services. This section provides a comprehensive overview of Event Grid's fundamentals, including its architecture, key components, and how it enables seamless communication between IoT devices and cloud services.

Event Grid architecture

Event Grid operates on a **publish-subscribe** (**pub-sub**) model, allowing for the decoupling of event producers and consumers. This architecture ensures that events generated by IoT devices can be efficiently delivered to interested parties without these devices needing to be aware of who or what is consuming their data. Here are the key components of Event Grid's architecture:

- **Event publishers**: These are the sources of events, which can be IoT devices, applications, or services. They generate events and publish them to Event Grid.

- **Event topics**: Event topics serve as the communication channels for events. They act as logical endpoints to which events are published. Each event topic can have multiple subscribers interested in specific event types.

- **Event subscribers**: Subscribers are entities that want to receive and act upon specific events. Subscribers can be Azure Functions, Logic Apps, Webhooks, or other custom endpoints capable of processing events.

- **Event domains**: Event domains group related event topics and provide a hierarchical structure for organizing and managing events. They simplify event routing and access control.

The following diagram shows a generic pub/sub architecture. The pub/sub architecture offers a number of advantages, discussed next:

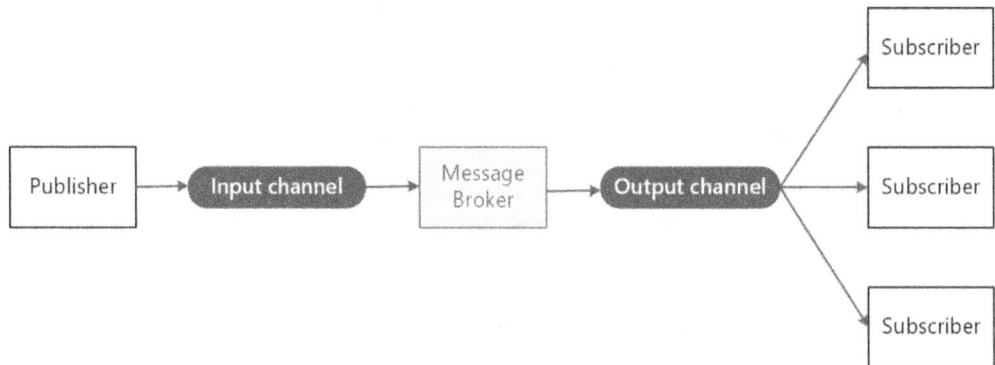

Figure 12.1 – Pub-sub architecture

Pub/sub messaging offers a range of advantages:

- **Decoupling of subsystems**: Pub/sub messaging untangles interconnected subsystems, allowing them to operate independently. Messages can be efficiently managed, even when one or more recipients are offline.

- **Enhanced scalability and sender responsiveness**: The sender can rapidly dispatch a single message to the input channel and then resume its core processing duties. The messaging infrastructure takes charge of ensuring that messages reach their intended recipients, bolstering scalability and sender responsiveness.

- **Improved reliability**: Asynchronous messaging fosters smooth application operation under increased workloads and equips systems to handle intermittent failures with greater resilience.

- **Deferred or scheduled processing**: Subscribers have the flexibility to delay message retrieval until off-peak hours or adhere to specific processing schedules, providing valuable control over workflow timing.

- **Simplified integration**: Pub/sub messaging facilitates seamless integration between systems with varying platforms, programming languages, or communication protocols. It also streamlines the connection between on-premises systems and cloud-based applications.

- **Support for asynchronous workflows**: The architecture enables the establishment of asynchronous workflows across an entire enterprise, promoting efficient data flow between disparate components.

- **Enhanced testability**: Monitoring channels and inspecting or logging messages becomes a seamless part of an overarching integration testing strategy, enhancing the ability to test and validate system behavior.

- **Separation of concerns (SoC)**: Each application can concentrate on its core functionality, leaving the messaging infrastructure to handle the intricate task of reliably routing messages to multiple consumers, thus fostering a cleaner SoC within your applications.

Event Grid key concepts

To fully grasp Event Grid fundamentals, it's crucial to understand the following key concepts:

- **Event**: An event represents a significant occurrence or change in state that is relevant to the IoT ecosystem. Examples of events in IoT include sensor data updates, device status changes, alarms, and more.

- **Event schema**: Events often have defined schemas that specify the structure and data associated with a particular event type. This schema ensures that both event producers and consumers understand the event's content.

- **Event subscription**: An event subscription defines the criteria for delivering events to subscribers. It includes information such as the event types to filter on, the endpoint to deliver events to, and authentication settings.

- **Dead lettering**: Event Grid provides mechanisms for handling events that fail to be delivered to subscribers. These undelivered events are stored in a dead-lettering system for analysis and troubleshooting.

- **Event routing and filtering**: Event Grid offers powerful routing and filtering capabilities, ensuring that events reach the right subscribers efficiently. You can define filters based on event type, subject, and other custom attributes, allowing you to precisely control which events are delivered to specific subscribers.

- **Event Grid and IoT**: For IoT scenarios, Event Grid plays a vital role in connecting the physical world of devices to the digital world of cloud services. It enables real-time communication between IoT devices and cloud applications, facilitating data processing, analysis, and decision-making.

In the following sections of this chapter, we'll explore how to create and manage Event Grid resources, set up event subscriptions, and respond to events effectively. Armed with these fundamentals, you'll be well prepared to harness the power of Event Grid in your IoT projects, enabling seamless communication and event-driven workflows.

Exploring common IoT events

In an IoT environment, a wide range of events can be sent to an event grid to enable real-time communication, data processing, and automation. Here are some common IoT events that are typically sent to an event grid:

- **Sensor data updates**: This is one of the most common types of IoT events. It includes data from various sensors such as temperature, humidity, light, motion, and more. These updates can trigger actions such as alerts, data storage, or analytics.

- **Device status changes**: Events indicating changes in the status of IoT devices, such as device startup, shutdown, reboot, or connectivity status changes. These events can help in monitoring and managing the health of IoT devices.

- **Alarm and alert events**: IoT devices often generate events when certain thresholds or conditions are met. For example, a smoke detector might send an event when it detects smoke, or a security camera might trigger an event when it detects motion.

- **Device firmware updates**: Events related to the updating of device firmware or software. This is crucial for maintaining the security and functionality of IoT devices.

- **Command and control events**: Events that carry commands from central systems or users to IoT devices, such as turning on/off lights, adjusting thermostat settings, or locking/unlocking smart doors.

- **Geolocation events**: Events that provide location information of IoT devices. These can be used for tracking assets and vehicles or monitoring the movement of people or objects.

- **Environmental events**: Events related to changes in the environment, such as air quality, pollution levels, noise levels, or weather conditions. These events can be used for environmental monitoring and control.

- **Energy consumption events**: For energy management systems, events related to energy usage, power outages, voltage fluctuations, or abnormal energy consumption patterns are vital.

- **Health and wellness events**: In healthcare IoT, events can include patient vital signs, medication reminders, or alerts for medical professionals in case of emergencies.

- **Inventory and supply chain events**: For logistics and supply chain management, events related to inventory levels, shipment tracking, and product expiration dates can be sent to Event Grid.

- **Security events**: Events indicating security breaches, unauthorized access attempts, or suspicious activities within an IoT network or system.

- **Device battery status**: For battery-powered devices, events related to battery levels, charging status, and low-battery warnings are essential for device management.

- **Predictive maintenance events**: These events are generated based on predictive analytics and sensor data to schedule maintenance before equipment failure occurs, optimizing operational efficiency.

- **Custom application events**: Events specific to the unique functionality of IoT devices or applications. These can vary widely depending on the use case and industry.

- **Device registration and de-registration**: Events that track when IoT devices are added or removed from the network.

In addition to these traditional events we can monitor with an event grid, Event Grid has recently added support for the MQTT protocol, which opens up a whole new category of use cases:

- **Efficient telemetry ingestion**: Utilize the many-to-one messaging pattern to seamlessly absorb telemetry data. This approach alleviates the need for your application to manage a multitude of device connections directly, shifting this responsibility to Event Grid.

- **Client control with request-response messaging**: Implement one-to-one messaging to exercise control over your MQTT clients. This approach empowers any client to engage in unrestricted communication with any other client, regardless of their respective roles or responsibilities.

- **Alert dissemination to multiple clients**: Employ the one-to-many messaging pattern to disseminate alerts to a fleet of clients. With this method, your application needs to transmit just one message, which Event Grid subsequently replicates and delivers to every interested client.

- **Seamless integration with Azure services**: Leverage Event Grid's ability to route MQTT messages to Azure services and Webhooks through HTTP push delivery. This integration seamlessly initiates data pipelines, commencing with the ingestion of data from your IoT devices, and allows you to harness the full power of Azure's capabilities for data processing and analysis.

Now that we have seen some common use cases and types of events we look for in an IoT system, it is time to look at how to send these events to an event hub for distribution.

Sending IoT events to Event Grid

One thing to remember is that Event Grid is used for any type of event-based processing in Azure. It can facilitate events from many different Azure services. Some common services are *storage*, *functions*, and *web applications*.

Azure IoT Hub seamlessly integrates with Azure Event Grid, allowing you to effortlessly dispatch event notifications to various services and initiate subsequent workflows. By configuring your business applications to actively monitor IoT Hub events, you can respond to pivotal occurrences with utmost reliability, scalability, and security. For instance, envision the creation of an application that automatically updates a database, generates a work ticket, and dispatches email notifications whenever a new IoT device is registered within your IoT hub.

To set up an event, navigate to the service providing the event (in our case, IoT Hub) and select the **Events** option in the menu:

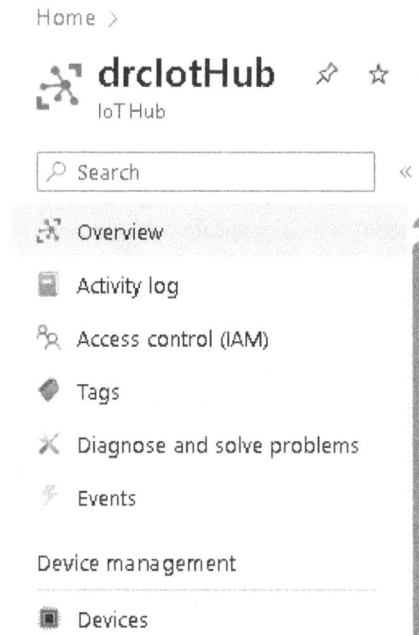

Figure 12.2 – Setting events

Once you select **Events**, you then create an event subscription. The types of events supported depend on the service issuing the event. For an IoT hub, the event subscription allows you to pick the events supported using a checkbox list, such as devices created, deleted, connected, and disconnected. It also supports telemetry. You also need to indicate the schema, which you can leave as the default Event Grid schema.

Next, you need to provide a topic. Event Grid enables you to react to events from various Azure services and custom sources by routing those events to different subscribers or handlers. The Event Grid namespace topic plays a crucial role in this EDA.

Here's what an Event Grid namespace topic is used for:

- **Event ingestion**: Event Grid namespaces are the central entry point for events. Events from various sources are sent to the namespace topic. These sources can include Azure services such as Azure Storage, Azure Blob Storage, Azure Functions, Azure IoT Hub, and more, as well as custom applications and third-party services.

- **Event routing**: The namespace topic is responsible for routing incoming events to subscribers based on event subscriptions. Subscribers can be various Azure resources, such as Azure Functions, Azure Logic Apps, Azure Event Hubs, Azure Automation runbooks, or custom Webhook endpoints. Event Grid ensures that events are sent to the appropriate subscribers based on filtering and routing rules defined in event subscriptions.

- **Fan-out and filtering**: Event Grid allows you to create event subscriptions that filter events based on specific criteria. This enables you to fan out events to multiple subscribers or selectively route events to specific endpoints. You can use the Event Grid namespace topic to manage and configure these event subscriptions.

- **Scalability and reliability**: Event Grid namespaces are designed to be highly scalable and reliable. They can handle a large volume of events and ensure that events are delivered in a timely manner to subscribers. Azure manages the underlying infrastructure, so you don't have to worry about scalability or availability concerns.

In summary, an Event Grid namespace topic serves as the central hub for receiving and routing events in Azure Event Grid. It allows you to decouple event producers from event consumers, enabling a highly scalable and flexible EDA in Azure.

Responding to events

The next step is to define an endpoint, which is the event handler that is responsible for processing an event. Currently, you can select quite a few handlers depending on how you plan on reacting to the event, as seen here:

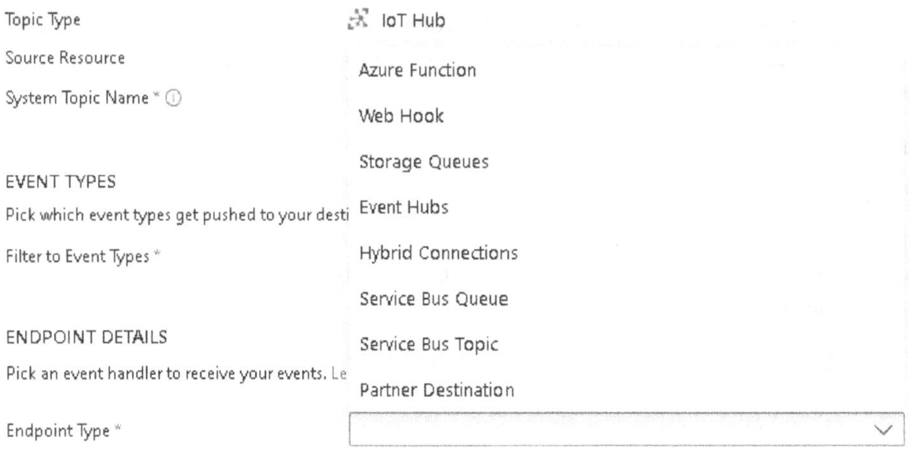

Topic Type	:X: IoT Hub
Source Resource	Azure Function
System Topic Name * ⓘ	Web Hook
	Storage Queues
EVENT TYPES	
Pick which event types get pushed to your desti	Event Hubs
Filter to Event Types *	Hybrid Connections
	Service Bus Queue
ENDPOINT DETAILS	Service Bus Topic
Pick an event handler to receive your events. Le	Partner Destination
Endpoint Type *	⌄

Figure 12.3 – Selecting an endpoint

You can also put filtering on your events. Filtering events from an Azure IoT hub can be a valuable tool for managing and processing IoT data efficiently. Here are some reasons why you might want to use filtering for IoT Hub events:

- **Reducing data volume**: IoT devices can generate a massive amount of data, and not all of it may be relevant to your specific use case. Filtering allows you to discard or route only the events that meet certain criteria, reducing the volume of data that needs to be processed and stored. This can help save on data storage costs and processing resources.

- **Selective routing**: Filtering enables you to route specific types of IoT events to different downstream services or endpoints based on their content or metadata. For example, you might want to send temperature sensor readings to one Azure function for real-time monitoring and humidity sensor readings to another for historical data analysis.

- **Security and access control**: You can use filtering to restrict access to sensitive IoT data. By filtering events and allowing only authorized subscribers to receive specific types of events, you can enhance security and ensure that sensitive data is only accessible to those with the proper permissions.

- **Custom processing**: Depending on the type of IoT events you're dealing with, you may want to apply custom processing logic to certain events. Filtering allows you to identify events that require specialized handling and route them to the appropriate processing pipeline.

- **Aggregation and summarization**: Filtering can be used to aggregate or summarize data from IoT devices. For example, you might filter and aggregate data from multiple devices to compute average values or detect anomalies, sending the aggregated results to a downstream analytics service.

- **Resource optimization**: By filtering out irrelevant events, you can optimize the resource usage of downstream services. For instance, you can avoid invoking expensive Azure functions or other compute resources for events that don't require processing, thereby saving on costs.

- **Compliance and reporting**: Filtering can help you meet compliance requirements by ensuring that only events meeting specific criteria are logged or reported. This can simplify compliance auditing and reporting processes.

- **Traffic shaping**: If your IoT solution needs to handle bursty traffic or manage traffic patterns, you can use filtering to control the flow of events to downstream systems. For instance, you might implement rate limiting or buffering based on filtered criteria:

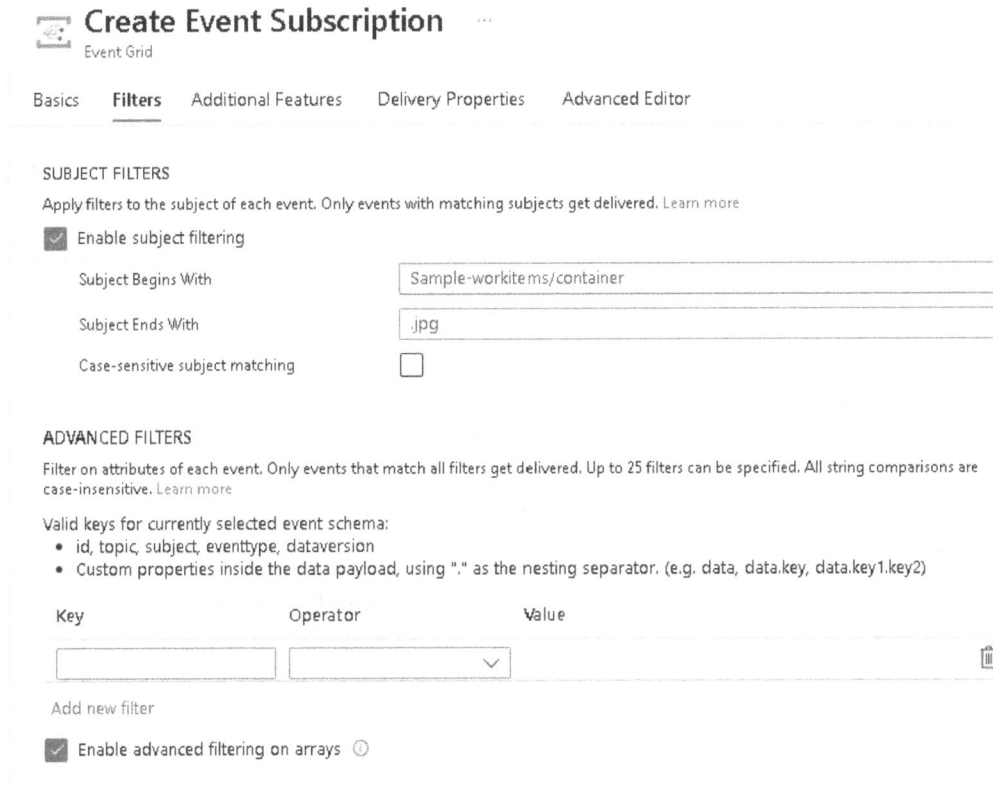

Create Event Subscription
Event Grid

Basics   **Filters**   Additional Features   Delivery Properties   Advanced Editor

SUBJECT FILTERS

Apply filters to the subject of each event. Only events with matching subjects get delivered. Learn more

☑ Enable subject filtering

Subject Begins With	Sample-workitems/container
Subject Ends With	.jpg
Case-sensitive subject matching	☐

ADVANCED FILTERS

Filter on attributes of each event. Only events that match all filters get delivered. Up to 25 filters can be specified. All string comparisons are case-insensitive. Learn more

Valid keys for currently selected event schema:
- id, topic, subject, eventtype, dataversion
- Custom properties inside the data payload, using "." as the nesting separator. (e.g. data, data.key, data.key1.key2)

Key	Operator	Value	
	∨		🗑

Add new filter

☑ Enable advanced filtering on arrays ⓘ

Figure 12.4 – Setting up event filtering

In summary, filtering for IoT Hub events allows you to efficiently manage and process the vast amount of data generated by IoT devices, ensuring that only relevant events are sent to downstream services, improving resource utilization, enhancing security, and enabling custom processing and routing based on your specific IoT use case and requirements.

Now that you have seen how to publish events, it is time to look at how you can subscribe and respond to events.

Subscribing and responding to IoT Hub events

Responding to IoT Hub events typically involves creating event handlers or subscribers that can process events generated by IoT devices and take appropriate actions. Here is some guidance on how to respond to IoT Hub events:

- **Create event handlers**:

  - Decide how you want to respond to IoT Hub events. Common response actions include storing data, sending alerts, performing analytics, or triggering downstream processes.

  - Choose Azure services or custom applications that will act as event handlers or subscribers for IoT Hub events. Some common choices include Azure Functions, Azure Logic Apps, **Azure Stream Analytics (ASA)**, and custom applications.

- **Azure Functions**:

  - If you're using Azure Functions as your event handler, create a new Azure function or use an existing one.

  - Configure an Event Hubs trigger or IoT Hub trigger in your Azure function to listen for events from your IoT hub.

  - Implement the desired processing logic in your Azure function. This can include data validation, transformation, storage, or any custom actions you want to take in response to the events.

- **Azure Logic Apps**:

  - If you prefer using Azure Logic Apps, create a new logic app.

  - Add an IoT Hub trigger to your logic app and configure it to receive events from your IoT hub.

  - Build your workflow in the logic app, including actions you want to perform in response to incoming IoT events. Logic Apps offers a wide range of built-in connectors for integration with other Azure services and external systems.

- **ASA**:

  - If real-time event processing and analytics are your focus, set up an ASA job.

  - Configure an input source to connect to your IoT hub.

  - Define queries in ASA to filter, transform, or aggregate incoming data.

  - Configure an output sink to send the processed data to the desired destination, such as Azure Storage, Azure Cosmos DB, or another service.

- **Custom applications**:

  - If you're building custom event handlers, implement a solution that can connect to the IoT hub using one of the supported protocols (for example, MQTT, AMQP, or HTTPS).

  - Create event processing logic within your application to consume IoT Hub events, perform necessary operations, and respond accordingly.

- **Event subscriptions**:

  - Configure event subscriptions in your IoT hub to specify which events should be routed to your chosen event handlers.

  - Define routing rules based on message properties, message types, or other criteria to ensure that only relevant events are sent to your handlers.

- **Testing and monitoring**:

  - Test your event-handling logic to ensure that it works as expected.

  - Implement monitoring and logging to track the performance and behavior of your event handlers and to diagnose any issues that may arise.

- **Scaling**: Depending on your requirements, you may need to scale your event handling infrastructure to handle high event volumes efficiently. Azure provides scaling options for Azure Functions, Logic Apps, and ASA to accommodate increased workloads.

- **Error handling and retry**: Implement error-handling and retry mechanisms in your event handlers to handle transient failures and ensure robust event processing.

By following these recommendations, you can effectively respond to IoT Hub events and build a responsive, scalable, and reliable IoT solution that meets your specific use case and requirements.

After creating the subscription, you will get some automatic insight into how it is performing:

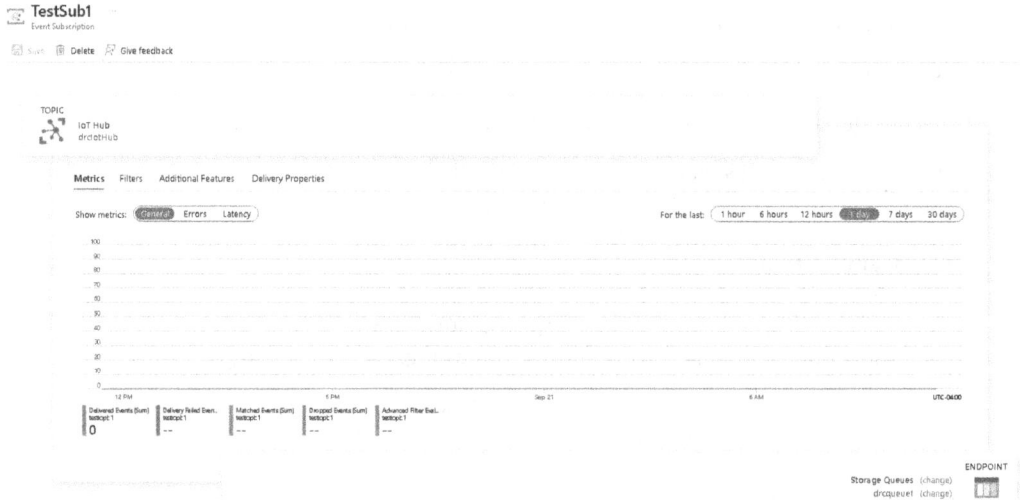

Figure 12.5 – Monitoring event performance

Now that we have discussed how IoT Hub and Event Grid can be integrated to provide automated responses or alerts to events, it is time to roll up your sleeves and complete the following lab.

Lab – monitoring device connection and disconnection with Event Grid

In this lab, we will create an automated email when a device connects or disconnects from an IoT hub:

1. In a new resource group, create an IoT hub.
2. On the menu for the IoT hub, select the **Events** option.

On the **Events** page, you can see some examples of event consumers:

Azure Event Grid natively supports these resources as event handlers. Learn more

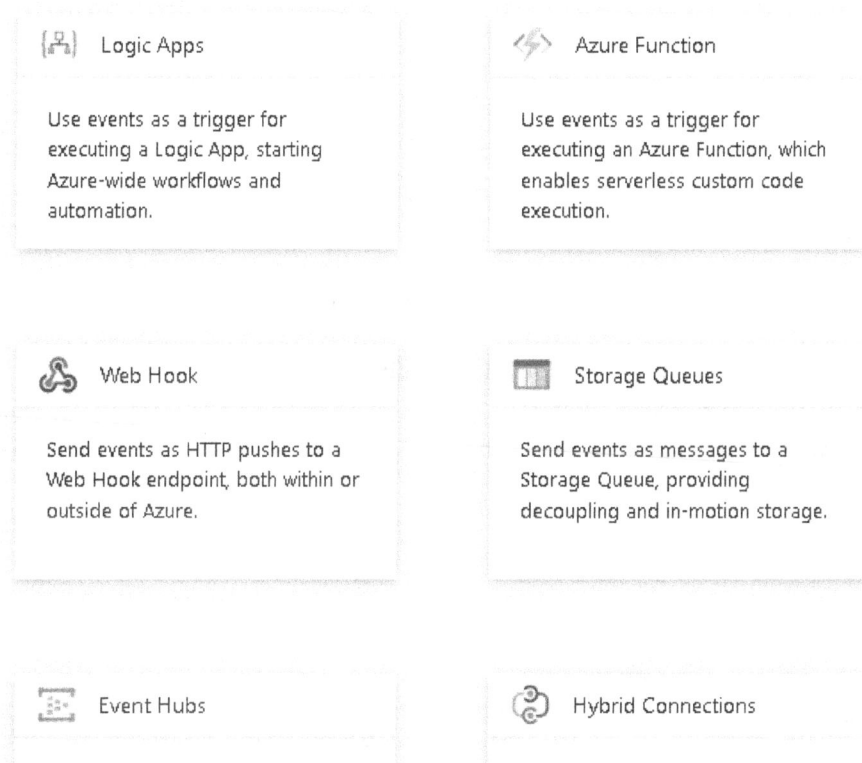

Logic Apps

Use events as a trigger for executing a Logic App, starting Azure-wide workflows and automation.

Azure Function

Use events as a trigger for executing an Azure Function, which enables serverless custom code execution.

Web Hook

Send events as HTTP pushes to a Web Hook endpoint, both within or outside of Azure.

Storage Queues

Send events as messages to a Storage Queue, providing decoupling and in-motion storage.

Event Hubs

Hybrid Connections

Figure 12.6 – Natively supported event handlers

3. Select the **Logic Apps** option. This will open the **Logic Apps Designer** page with a connection to the Event Grid service:

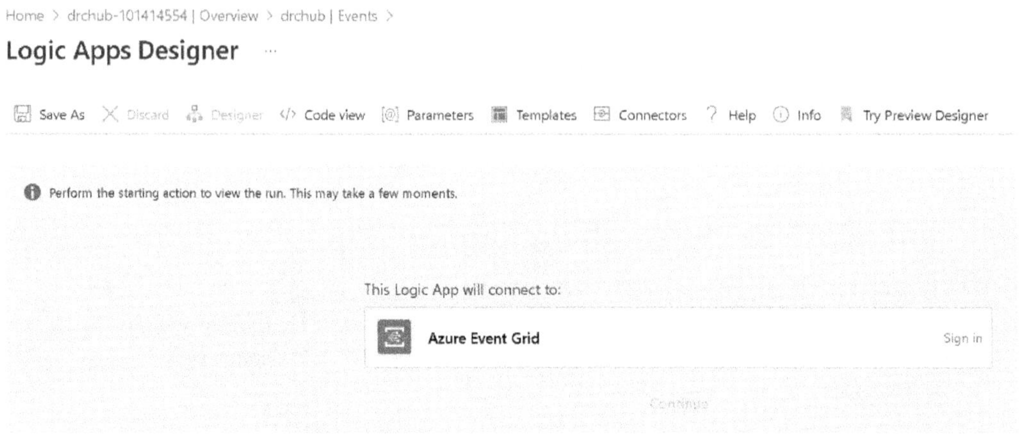

Home > drchub-101414554 | Overview > drchub | Events >

Logic Apps Designer ...

Save As   Discard   Designer   </> Code view   Parameters   Templates   Connectors   ? Help   Info   Try Preview Designer

ℹ Perform the starting action to view the run. This may take a few moments.

This Logic App will connect to:

Azure Event Grid   Sign in

Continue

Figure 12.7 – Creating a logic app to respond to the event

4. Sign in to the Event Grid service. Select the **Continue** button after signing in.

5. Select **Microsoft.Devices.IoTHubs** as the resource type and select **Microsoft.Devices. DevicesConnected** as the event type. Enter a value (of your choice) for the **Subscription** name:

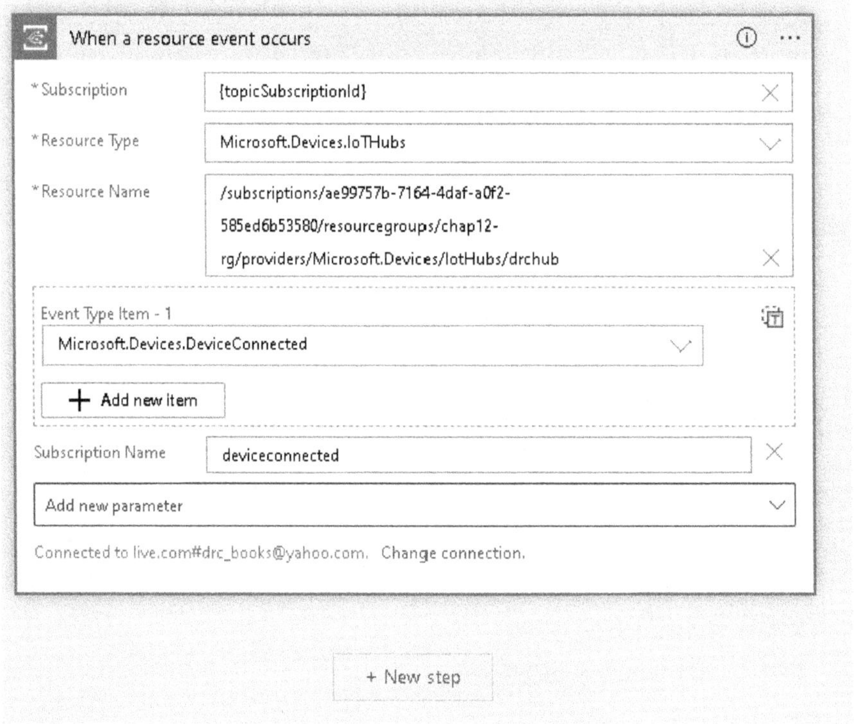

When a resource event occurs   ℹ ...

* Subscription   {topicSubscriptionId}   ✕

* Resource Type   Microsoft.Devices.IoTHubs   ⌄

* Resource Name   /subscriptions/ae99757b-7164-4daf-a0f2-
                      585ed6b53580/resourcegroups/chap12-
                      rg/providers/Microsoft.Devices/IotHubs/drchub   ✕

Event Type Item - 1
   Microsoft.Devices.DeviceConnected   ⌄

   + Add new item

Subscription Name   deviceconnected   ✕

Add new parameter   ⌄

Connected to live.com#drc_books@yahoo.com. Change connection.

+ New step

Figure 12.8 – Filling out the necessary values

6. Select the + **New step** option, and on the next page, select the option to send an email from your Office 365 account.

7. A sign-in popup will appear. Sign in to your account and send yourself an email.

8. Click on the **Subject** box and add the subject of the mail, for example, `Test Response`.

9. Click on the **Body** field and add the body text of the message passed in:

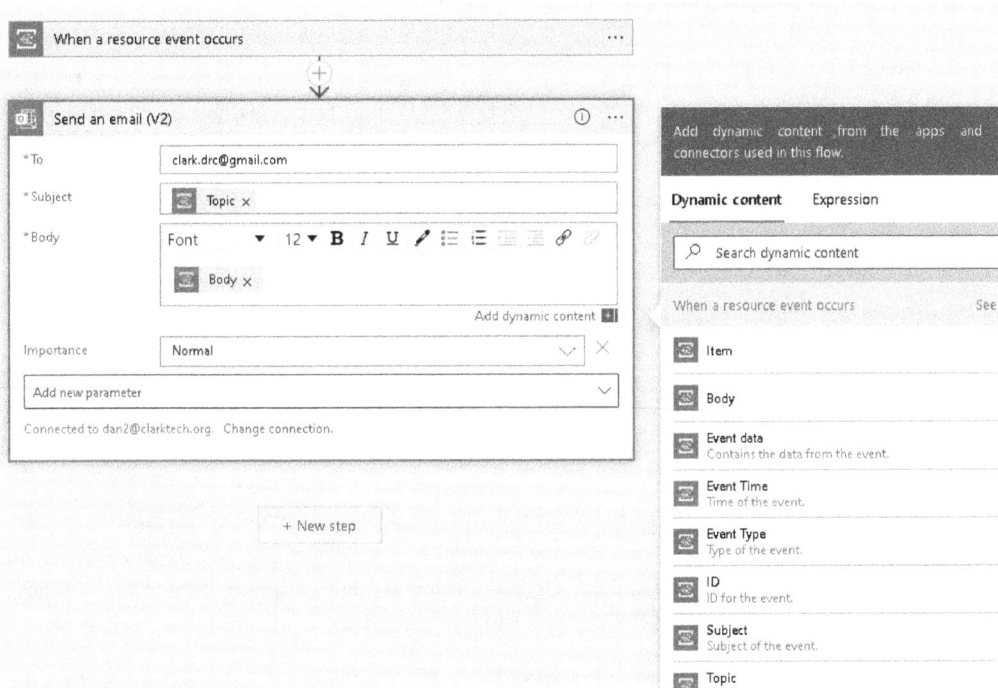

Figure 12.9 – Configuring the email

10. Save and create the logic app by selecting the **Save as** icon at the top of the designer.

11. Navigate back to the resource group you created for this lab. You should now see the logic app and an Event Grid system topic added to the resource group:

Figure 12.10 – Viewing the new resources added to the resource group

12. To test the event, add a device to the IoT hub. Copy the connection string and paste it on *line 15* of the device simulator found here: `https://azure-samples.github.io/raspberry-pi-web-simulator/`.

13. Run the simulator. After a few minutes, you should see an email arrive with a body that looks similar to the following:

```
{"id":"b4115eb6-8070-d02f-6d1a-56738e624b61","topic":"/
SUBSCRIPTIONS/AE99757B-7164-4DAF-A0F2-585ED6B53580/RESOURCEGROUPS/
CHAP12-RG/PROVIDERS/MICROSOFT.DEVICES/IOTHUBS/
DRCHUB","subject":"devices/device01","eventType":"Microsoft.
Devices.DeviceConnected","data":{"deviceConnectionStateEventInfo":
{"sequenceNumber":"000000000000000001D9ECB9250CBF2B
0000000200000000000000000000000001"},
"hubName":"drchub","deviceId":"device01"},"dataVersion":"
","metadataVersion":"1","eventTime":"2023-09-22T15:08:52.975146Z"}
```

14. After verifying that you are getting emails, clean up your resources by deleting the resource group.

Now that you have seen how you can react to events in your IoT system, let's review what you have learned.

Summary

In this chapter, we embarked on a journey through the intricacies of Event Grid, focusing on its pivotal role in managing IoT events effectively. We began by establishing a solid foundation in the fundamentals of Event Grid, gaining insight into its architecture, key components, and the seamless communication it facilitates between IoT devices and cloud services.

Next, we delved into the rich use cases of common IoT events, uncovering the diverse range of events that populate the IoT landscape. These events, from sensor data updates to device status changes, form the lifeblood of IoT systems and lay the groundwork for responsive, intelligent applications.

Our exploration continued as we ventured into subscribing to events. Here, we learned how to create event subscriptions, configure event filters, and choose the appropriate delivery methods to ensure that IoT events are routed precisely where they are needed. With this knowledge, you're now equipped to fine-tune event management for your IoT solutions.

Finally, you learned how to respond to events, design event-driven workflows, automate actions, and orchestrate processes triggered by specific IoT events. By mastering the art of event response, you have the power to transform data into actionable insights, enabling your IoT ecosystem to thrive.

Armed with these Event Grid fundamentals, an understanding of common IoT events, and the ability to subscribe to and respond to events effectively, you are well prepared to navigate the evolving landscape of IoT event management. Whether you're building intelligent IoT applications, optimizing device connectivity, or enhancing your event-driven workflows, this chapter has equipped you with the knowledge and tools to succeed in the dynamic world of IoT.

This chapter brings us to the end of our journey to mastering setting up IoT systems on Azure. As with everything in technology, it is always in a state of flux. Although interfaces and tools may change, if you have a strong grasp of the fundamentals, you will easily evolve with the change. I hope you have enjoyed the journey, and good luck with your future endeavors in the IoT landscape!

Index

A

‹packt›

Packtpub.com

Subscribe to our online digital library for full access to over 7,000 books and videos, as well as industry leading tools to help you plan your personal development and advance your career. For more information, please visit our website.

Why subscribe?

- Spend less time learning and more time coding with practical eBooks and Videos from over 4,000 industry professionals

- Improve your learning with Skill Plans built especially for you

- Get a free eBook or video every month

- Fully searchable for easy access to vital information

- Copy and paste, print, and bookmark content

Did you know that Packt offers eBook versions of every book published, with PDF and ePub files available? You can upgrade to the eBook version at packtpub.com and as a print book customer, you are entitled to a discount on the eBook copy. Get in touch with us at customercare@packtpub.com for more details.

At www.packtpub.com, you can also read a collection of free technical articles, sign up for a range of free newsletters, and receive exclusive discounts and offers on Packt books and eBooks.

Other Books You May Enjoy

If you enjoyed this book, you may be interested in these other books by Packt:

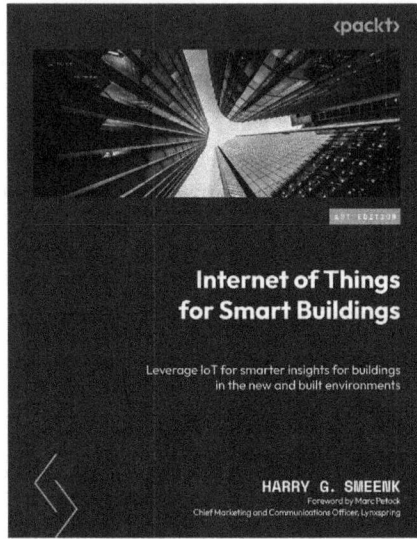

Internet of Things for Smart Buildings

Harry G. Smeenk

ISBN: 9781804619865

- Discover what's a smart building and how IoT enables smart solutions
- Uncover how IoT can make mechanical and electrical systems smart
- Understand how IoT improves workflow tasks, operations, and maintenance
- Explore the components and technology that make a smart building
- Recognize how to put together components to deploy smart applications
- Build your smart building stack to design and develop smart solutions

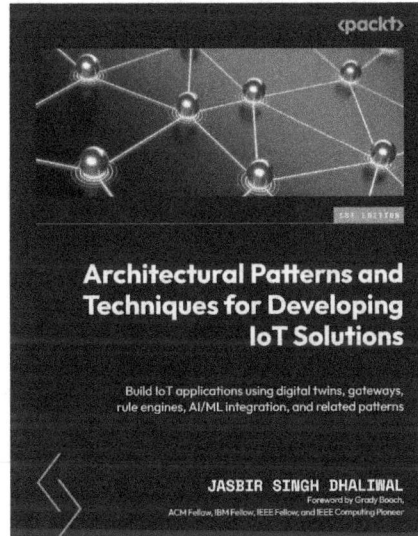

Architectural Patterns and Techniques for Developing IoT Solutions

Jasbir Singh Dhaliwal

ISBN: 9781803245492

- Get to grips with the essentials of different architectural patterns and anti-patterns
- Discover the underlying commonalities in diverse IoT applications
- Combine patterns from physical and virtual realms to develop innovative applications
- Choose the right set of sensors and actuators for your solution
- Explore analytics-related tools and techniques such as TinyML and sensor fusion
- Overcome the challenges faced in securing IoT systems
- Leverage use cases based on edge computing and emerging technologies such as 3D printing, 5G, generative AI, and LLMs

Packt is searching for authors like you

If you're interested in becoming an author for Packt, please visit `authors.packtpub.com` and apply today. We have worked with thousands of developers and tech professionals, just like you, to help them share their insight with the global tech community. You can make a general application, apply for a specific hot topic that we are recruiting an author for, or submit your own idea.

Share Your Thoughts

Now you've finished *The Azure IoT Handbook*, we'd love to hear your thoughts! Scan the QR code below to go straight to the Amazon review page for this book and share your feedback or leave a review on the site that you purchased it from.

`https://packt.link/r/1837633614`

Your review is important to us and the tech community and will help us make sure we're delivering excellent quality content.

Download a free PDF copy of this book

Thanks for purchasing this book!

Do you like to read on the go but are unable to carry your print books everywhere? Is your eBook purchase not compatible with the device of your choice?

Don't worry, now with every Packt book you get a DRM-free PDF version of that book at no cost.

Read anywhere, any place, on any device. Search, copy, and paste code from your favorite technical books directly into your application.

The perks don't stop there, you can get exclusive access to discounts, newsletters, and great free content in your inbox daily

Follow these simple steps to get the benefits:

1. Scan the QR code or visit the link below

https://packt.link/free-ebook/978-1-83763-361-6

2. Submit your proof of purchase
3. That's it! We'll send your free PDF and other benefits to your email directly

www.ingramcontent.com/pod-product-compliance
Lightning Source LLC
Chambersburg PA
CBHW061810210326
41599CB00034B/6946